The Gene Revolution

The Gene Revolution
GM Crops and Unequal Development

Edited by
Sakiko Fukuda-Parr

London • Sterling, VA

First published by Earthscan in the UK and USA in 2007

Copyright © Sakiko Fukuda-Parr, 2007

ISBN: 978-1-84407-409-9   paperback
        978-1-84407-410-5   hardback

Typesetting by Safehouse Creative
Printed and bound in the UK by Bath Press, Bath
Cover design by Susanne Harris

For a full list of publications please contact:

Earthscan
8–12 Camden High Street
London, NW1 0JH, UK
Tel: +44 (0)20 7387 8558
Fax: +44 (0)20 7387 8998
Email: earthinfo@earthscan.co.uk
Web: www.earthscan.co.uk

22883 Quicksilver Drive, Sterling, VA 20166-2012, USA

Earthscan is an imprint of James and James (Science Publishers) Ltd and
publishes in association with the International Institute for Environment
and Development

A catalogue record for this book is available from the British Library

Library of Congress Cataloging-in-Publication Data has been applied for

The paper used for the text of this book is FSC certified
FSC (The Forestry Stewardship Council) is an
international network to promote responsible
management of the world's forests.

Mixed Sources
Product group from well-managed forests and other controlled sources
www.fsc.org Cert no. SGS-COC-2121
© 1996 Forest Stewardship Council

Printed on totally chlorine-free paper

Contents

List of Figures, Tables and Boxes

Figures

Tables

Boxes

List of Contributors

Editor

Sakiko Fukuda-Parr is visiting professor at the New School University, New York. She developed this volume as research fellow at the Kennedy School of Government, Harvard University. From 1995 to 2004 she was director and chief author of the United Nations Development Programme (UNDP)'s annual publication that advocates people-centred development: *Human Development Report*. The ten reports she has directed cover diverse topics from poverty to growth to human rights to democracy, including the 2001 report, *Making New Technologies Work for Human Development*. Her other publications include *Readings in Human Development*; *Capacity for Development: Old Problems, New Solutions*; and *Rethinking Technical Cooperation, Reforms for Capacity Building in Africa*, as well as numerous articles. She has had a long career in international development at the UNDP and the World Bank.

Her current research interests focus on several aspects of international political economy – technology and institutions, human rights, and development aid. She is a citizen of Japan and a graduate of the University of Cambridge, the Fletcher School of Law and Diplomacy and the University of Sussex.

Contributors

Walter S. Alhassan is coordinator of the Programme for Biosafety Systems (PBS) for West and Central Africa, currently covering Ghana, Mali and Nigeria. PBS is directed by the International Food Policy Research Institute (IFPRA) and hosted in West Africa by the Forum for Agricultural Research in Africa (FARA) based in Accra, Ghana.

In 2000 and 2002 Professor Alhassan conducted and published detailed studies on the status and application of agro-biotechnology in a cross-section of West and Central African countries within the West and Central African Council for Agricultural Research and Development (CORAF/WECARD) area. In 2004, he coordinated the development of proposals for strengthening the 21 countries in West and Central Africa under CORAF/WECARD in biotechnology and biosafety. More recently, he has been West and Central African Coordinator of the Agricultural Biotechnology Support Program (ABSP), Phase II, Director-General of the Ghana Council for Scientific and Industrial Research and, before that, professor of animal science and dean of the School of Agriculture at the Abubakar Tafawa Balewa University in Bauchi, Nigeria. Professor Alhassan's extensive publications on animal agriculture have covered the production and utilization of crop residues and the general area of livestock–crop integration.

Izaías de Carvalho Borges holds a BSc in economics from the Federal University of Viçosa (Minas Gerais) and is a PhD candidate in economics at the University of Campinas and a researcher at the Center of Agricultural Economics, University of Campinas (IE/UNICAMP). Since 2005 he has taught economics at the Center of Economics and Administration at the Catholic University of Campinas.

His areas of research are economics of innovation, biotechnology and multi-criteria methods.

Daniel Chudnovsky, an Argentinian economist who holds a DPhil from Oxford University, is full professor of international business and development economics at the University of San Andrés and Director of the Centro de Investigaciones para la Transformación (CENIT) in Argentina. He was formerly professor of development economics at the University of Buenos Aires, Director of the Center for International Economics in Buenos Aires and economic affairs officer at the United Nations Conference on Trade and Development (UNCTAD) in Geneva.

Professor Chudnovsky has directed several international research projects, written extensively on trade, foreign direct investment, environment and technology issues and published several books and papers in academic journals. His most recent books are: *¿Por qué sucedió? Las causas económicas de la reciente crisis Argentina* (with C. Bruno), *Siglo XXI de Argentina Editores*, Buenos Aires, 2003; and *Los transgénicos en la agricultura Argentina: Una historia con final abierto* (with E. Trigo, E. Cap and A. López), *Libros del Zorzal e IICA*, Buenos Aires, 2002.

José Maria Ferreira Jardim da Silveira, an agronomist with a PhD in economics, has been assistant professor at the Institute of Economics, IE/UNICAMP since 1985. He is also Director of the Center of Agriculture and Environmental Economics Studies, IE/UNICAMP and a senior researcher in the Study Group on the Organization of Research and Innovation/Institute of Geosciences at the University of Campinas; counsellor at the Biotechnology Information Center, Brazil (from 2003 till the present time); and a researcher, at the National Science and Research Council, level 2. His areas of interest are biotechnology (in particular property rights, venture capital, corporate finance, knowledge and innovation policies) and agriculture (in particular public programme evaluation methods, governance structure studies and mechanisms of governance design). He has published papers in several books and journals.

Marnus Gouse is a full-time researcher in the Department of Agricultural Economics, Extension and Rural Development at the University of Pretoria in South Africa, working towards his PhD in agricultural economics. He is currently leading a Rockefeller Foundation-supported research project studying the economic, socio-economic and health effects of small-scale farmer adoption of genetically modified (GM) white maize in South Africa.

Mr Gouse is also the project coordinator for the Food, Agricultural and Natural Resource Policy Analysis Network (FANRPAN) and PBS/US Agency for International Development (USAID)-supported Regional Approach to Biosafety for Southern African Countries (RABSAC) project. In 2000–2003 he was the main researcher in another Rockefeller Foundation-supported research project, which focused on socio-economic and farm-level impacts of GM crop adoption in South Africa – the first of its kind to concentrate on insect resistant maize and cotton adoption by commercial and small-scale farmers in Africa.

Jikun Huang, Founder and Director of the Centre for Chinese Agricultural Policy (CCAP) of the Chinese Academy of Sciences (CAS), as well as professor and chief scientist at China's Institute of Geographical Sciences and Natural Resources Research, is widely recognized as one of the country's leading agricultural economists. His research covers a wide range of issues on China's agricultural and rural economy, including agricultural research and development policy, resource and environmental economics, pricing and marketing, food consumption, poverty and trade liberalization. He has led more than 40 internationally and domestically funded research projects, published some 80 papers and articles in international journals (including *Science*, *Nature* and many in the field of economics), as well as over 130 papers in Chinese journals, and has authored eight books. He has received several awards from the Chinese Government, including one for being amongst the top ten outstanding young scientists (2002) and the Outstanding Scientific Progress award from the Ministry of Agriculture (three times).

Marcel Nwalozie is the Scientific Coordinator of CORAF/WECARD, an inter-state institution facilitating and coordinating agricultural research cooperation for development involving 22 countries. A crop physiologist with a PhD in field crops physiology (cowpea), and a versatile researcher, Dr Nwalozie has led several collaborative research projects on the adaptation of crops to drought. In the mid-1990s he was appointed international expert at the Regional Centre for the Study of Crop Adaptation to Drought (CERAAS), and has participated in several agricultural research priority identification, design and research proposal evaluations. He also led/supervised strategic planning and priority setting for agricultural research cooperation for West and Central Africa; and for the development of the West and Central African blueprint for biotechnology and biosafety.

Dr Nwalozie has also taught and supervised postgraduate students at Nigeria's Abia and Imo state universities, rising to senior lecturer status. He has published over 40 articles and chapters in books on crop physiology and research cooperation management.

Carl E. Pray is currently professor II and chairman of the Department of Agricultural, Food and Resource Economics, Rutgers, The State University of New Jersey. He is also an adjunct professor and advisor, PhD Programme, and a member of the Board of Academic Advisors of the Centre for Chinese Agricultural Policy, Chinese Academy of Science. He has been a visiting professor at Princeton Universities, Yale University and the University of Pennsylvania. The focus of Dr Pray's current research is agricultural science and technology policy in China, India and other developing countries. In the recent past he has studied how patents and industry concentration have affected biotech research in the US, how public policies could induce private companies to conduct research that would reduce hunger and poverty in developing countries, and the political and economic effects of public sector research in developing countries. The results of this research have been published in *Science*, *Nature*, the *American Journal of Agricultural Economics* and elsewhere.

Bharat Ramaswami is a professor of economics at the Planning Unit of the Indian Statistical Institute, Delhi. He earned his PhD from the University of Minnesota after receiving his Bachelor's and Master's degrees in economics from the University of Delhi. Issues prominent in his research are the design of safety nets, the supply of producer services in agricultural development and the pathways through which poor households gain from economic growth.

Harold Ransford Roy-Macauley, a native of Sierra Leone, is regional coordinator of the World Agroforestry Centre in West and Central Africa. He was previously a lecturer at the University of Sierra Leone, from which he was seconded as regional research scientist at CERAAS, a regional research and training unit of (CORAF/WECARD), based in Senegal. Later he was appointed Director of CERAAS and served in this position for nearly ten years. Dr Roy-Macauley has also been involved in coordination, planning and evaluation processes for the application of modern biotechnology to agricultural research, mainly in West and Central Africa.

During over 15 years of experience in scientific leadership, Dr Roy-Macauley has contributed to enhancing the scientific capacity of national agricultural research systems of developing countries – mainly in Africa. He has worked with public and private research, training and development institutions, farmers and civil society organizations at both Africa-wide and international levels. Most of his research has focused on plant biodiversity improvement and agricultural development in dry areas, resulting in numerous scientific publications.

Scott Rozelle is the Helen Farnsworth Senior Fellow and professor at Stanford University's Freeman Spogli Institute for International Studies and professor in the Department of Agricultural and Resource Economics at the University of California (UC), Davis. Professor Rozelle received his BSc from UC, Berkeley, and his MSc and PhD from Cornell University. Before moving to Stanford in 2006, he was (and remains) part of the faculty at the UC and is currently the UC Davis 2000 Chancellor Fellow – an award given each year to one of the university's outstanding faculty members. He has won the American Agricultural Economics Association (AAEA) Quality of Research Discovery Award twice (2003, 2005).

Professor Rozelle is fluent in Chinese and has established a research programme in which he has close working ties with several Chinese collaborators. In the past, his service to the AAEA has included being chair-elect of the International Section; associate editor, *Choices*; international committee member; and founding member of the Coordinating Committee for Research on China (WCC#101). He is the chair of the Board of Academic Advisors of the Center for Chinese Agricultural Policy.

Paco Sereme is executive secretary of CORAF/WECARD, charged with the day-to-day management of Council business including implementation of the decisions of its Governing Board and General Assembly. He was previously the director-general of the Institut de l'Environnement et de Recherches Agricoles (INERA) in Burkina Faso, where he had headed the Plant Pathology Laboratory, specialized in the seed pathology of pearl millet, sorghum, cowpea and bambara groundnut. At that time his focus was on the identification of indigenous natural products suitable for use in control of seed-borne diseases.

During his tenure at INERA, Dr Sereme supervised scores of young scientists from the University of Ouagadougou in their theses preparation. He has wide international cooperation experience with several centres of the Consultative Group on International Agricultural Research (CGIAR) and advanced research centres, both north and south. He has published widely in peer-reviewed journals.

Yves Tiberghien, assistant professor, Department of Political Science, University of British Columbia (Vancouver, Canada) is currently on leave as a Harvard Academy Scholar, where his research focuses on the politics of GMO regulations and the battle for global governance of biotechnology. This multi-year project is developing a typology of regulatory outcomes around the world, including in the European Union, Japan, Republic of Korea and China (see project website at www.gmopolitics.com).

Professor Tiberghien earned his PhD in political science from Stanford University in 2002. In 1988–1989 he spent 15 months at the Illinois Institute of Technology in Chicago studying genetic engineering, and wrote a thesis on recombinant *Bacillus* bacteria. He has published several articles and book chapters on Japanese political economy and corporate governance reforms. His most recent publication on GMOs is *The Battle for the Global Governance of Genetically Modified Organisms: The Roles of the European Union, Japan, Korea and*

China in a Comparative Context (2006) Les Etudes du CERI, no 124 (April), Institut d'Etudes Politiques, Paris. A book, *The Corporate Restructuring Dilemma: Global Incentives, Political Entrepreneurs, and Corporate Governance Reforms* is expected to be published in 2007.

Greg Traxler is a professor of agricultural economics at Auburn University, Alabama. A Minnesota native, Professor Traxler received a BBA from the College of Business Administration, University of Portland (Oregon); an MSc from the University of Minnesota, Department of Agricultural and Applied Economics, and his PhD from Iowa State University Department of Economics, in 1990. His research interests include the effects of technological change in agriculture in developed and developing countries, the impact of biotechnology research, the valuation of genetic plant resources and the distribution of benefits from patented crop varieties. He has completed studies on the diffusion and impact of *Bt* cotton in the US and in Mexico, and of RR soybean in the US and Argentina.

Professor Traxler has also worked on projects for the Food and Agriculture Organization of the UN (FAO), the Inter-American Development Bank, the Consultative Group on International Agricultural Research, The World Bank and the International Maize and Wheat Improvement Center.

Glossary of Commonly Used Terms

Abiotic – A nonliving substance or process.

Aflatoxin – A highly toxic and powerful carcinogen released by the fungus *Aspergillus flavus*, which can grow in improperly stored grain.

Ammonium Glufosinate – A broad-spectrum contact herbicide used to control a wide range of weeds. It acts by inhibiting the activity of the enzyme glutamine synthetase, thereby halting photosynthesis and killing the plant. Glufosinate inhibits the same enzyme in animals.

Arabidopsis – *Arabidopsis thaliana* is a genus of flowering plants found in northern temperate regions. It is often used in genetic experiments because it has a small genome, a rapid life cycle, and only grows 15 to 30cm in height. Related to brassicas, including rapeseed.

***Bemiscia tabaci* (also *Bemisia tabaci*)** – A sweet potato whitefly (sometimes known as a 'tobacco whitefly' or 'cotton whitefly') that can often cause damage to crops in tropical and sub-tropical areas.

Bioballistics – A method of transferring DNA. The foreign DNA is coated on extremely tiny metal slivers, which are shot into leaf discs using a 'shotgun'-like device. Once inside the cells of the leaf, some of the foreign DNA is taken up into the nucleus and incorporated into the host genome.

Bio-remediation – Using biological organisms to resolve an environmental problem, for example, using soil micro-organisms to break down an organic contaminant in the soil.

Biotic – Relating to life or living creatures.

Blackgram – *Vigna mungo*, a legume originating in India and cultivated in southern Asia.

Bromozynil (also Bromoxynilso) – An agricultural herbicide (trade name Buctril). The first genetically engineered variety of cotton to enter the US market was designed by Calgene to be resistant to bromozynil. The herbicide was subsequently banned by the Environmental Protection Agency (EPA).

***Bt* (*Bacillus thuringiensis*)** – A soil-living bacteria that produces an endotoxin deadly to insects. Many varieties exist, some of which target specific orders of insect. The genes encoding this endotoxin (referred to as *Bt* genes) are often inserted into genetically modified crops to give those crops a form of insect resistance.

Busseola fusca (or B. fusc) – African maize stem borer.

Cercospora – A fungus that often infects plants such as sugar beets and asparagus.

Chilo partellus – African grain stem borer.

Citrus variegated chlorosis (CVC) – A bacteria that infects citrus trees. It has adversely affected sweet orange crops in Argentina and Brazil.

Codex Alimentarius – Created in 1963 by FAO and the World Health Organization (WHO) to develop food standards, guidelines and codes of practice for application to consumer health, trade practices and coordination of international governmental and non-governmental organizations.

Edaphic – Of or relating to soil.

Edapho-climatic – Relating to both soil type and climate type.

Event – The successful transfer of a gene from one species to another.

Fusarium moniliforme – A type of fungus that commonly infects plants. It is capable of producing a variety of mycotoxins that can be dangerous to human health.

Geminivirus – A DNA virus with a small genome that infects plants. Geminiviruses rely on their host cell's replication processes and so have been used by scientists to study the regulation of DNA replication in plant cells. They are capable of infecting a wide variety of plant species and pose a serious threat to crops.

Genome – The full DNA sequence of an organism. The human genome contains approximately three billion chemical base pairs.

Genomics – Using the sequence of the entire DNA content of an organism (its genome) for the purpose of understanding the structure, function and expression patterns of its genes.

Germplasm – The genetic material that carries the inherited characteristics of an organism. In plant conservation, any part of the organism that acts as a vehicle for genetic material (i.e. seeds, fruit).

Glyphosphate – A common type of herbicide that inhibits a specific enzyme, EPSP synthase, which plants need to grow. Roundup, produced by Monsanto, is probably the most famous glyphosphate herbicide.

GM (genetic modification; also 'genetic engineering') – a process through which individual genes from one species of plant (or animal) are extracted and then inserted into the cell of another species.

Heartwater – An acute febrile disease of cattle, sheep and goats in sub-Saharan Africa. It is caused by a bacterium of the genus Cowdria and transmitted by ticks.

Helicoverpa armigera – The cotton bollworm, an insect pest capable of damaging a wide variety of crops.

Helminthiasis – A disease in which a part of the body is infested with worms such as tapeworms or roundworms.

Heterosis – Hybrid vigour. A situation in which breeding two plant lines results in progeny that are superior to both of the parent lines.

Imidazolinone – A type of herbicide (brand name Pursuit) that stops the activity of an enzyme, acetolactate synthase (ALS), vital for growth. Some varieties of maize have been bred to be resistant to Imidazolinone.

Lepidoptera – The second-largest order of insects, comprising 180,000 species of moths, butterflies and skippers.

Lignification – A change in the character of a plant's cell wall in which it becomes harder. Wood is created by the depositing of the organic substance, lignin, in a cell wall.

MDGs (Millennium Development Goals) – The Millennium Development Goals are a set of eight time-bound and measurable targets that are intended to provide a blueprint for the world's development efforts. The eight goals are: (1) eradicating extreme poverty and hunger; (2) achieving universal primary education; (3) promoting gender equality and empowering women; (4) reducing childhood mortality; (5) improving maternal health; (6) combating HIV/AIDS, malaria, and other diseases; (7) ensuring environmental sustainability; and (8) developing a global partnership for development.

Nematode – A roundworm that first infects the lungs and later the intestines. It usually enters the body through ingestion of contaminated food or contact with contaminated soil.

***Orysa* sp** – Relatives of wild rice that can be an important source of novel genes for genetic modification of rice.

Phaseolus vulgaris – Common bean, indigenous to South and Central America, including kidney, wax, cannellino and green beans.

Phytogenic – Having a plant origin.

Phytopathogen – An organism that causes a disease in a plant.

***Phytophtora* sp** – Root rot, a soil-borne fungal disease that can affect most plants.

PPP (purchasing power parity) – A method of more accurately comparing incomes across countries by taking into account differences in the cost of living. For example, if a certain basket of good costs 10 dollars in the US and 100 pesos in Mexico, then the PPP exchange rate would be 1 dollar = 10 pesos.

Proteomics – The study of proteins and their functions.

Pyriculariose – The French name for rice blast fungus, the disease caused by *Magnaporthe grisea*.

Rhizobium – A genus of bacteria that plays a very important role in agriculture by inducing nitrogen-fixing nodules on the rules of legumes such as peas, beans, clover and alfalfa. *Rhizobium* reduces the need for additional nitrogen fertilizers on these crops.

Rhizosphere – The soil zone immediately surrounding a plant's root system.

Sclerospora graminicola – The cause of downy mildew in pearl millet.

Spodoptera – A genus of arthropods that includes cutworm and armyworm.

Stacked genes – The insertion of multiple transgenes into an organism; for example, genes for resistance to herbicide and resistance to insects.

Stenocarpella maydis – A fungal pathogen that causes stem and ear rot of maize.

Striga – *Striga hermonthica* is a parasitic weed that affects food crops such as sorghum, maize, millet and rice. It is normally found in semi-arid and tropical parts of Africa.

Terminator genes – A colloquial name for transgenes developed by Delta & Pineland Co. (D&PL) that result in sterile seeds in the second generation. Seeds of the first generation can be planted but the seeds from the resulting plants are sterile and produce no crop.

Tospovirus – A subset of the Bunyaviridae virus family that can infect plants – including a range of ornamental and vegetable crops.

***Trait loci* (aka Quantitative Trait Locus or 'QTL')** – A region of DNA that is statistically associated with the expression of a particular trait in an organism. The trait in question is usually continuous and quantitative (e.g. height) rather than qualitative (e.g. colour) and therefore cannot be mapped to a single responsible gene.

Transgenes – Any gene that has been transferred using genetic modification from one species to another.

Transgenic – An organism that contains one or more genes from a different species.

Trypanosomiasis (aka 'sleeping sickness') – a disease endemic to sub-Saharan Africa caused by protozoa of the genus Trypanosoma and transmitted by the bite of a tsetse fly. Sleeping sickness can cause fever, weakness and sometimes death.

Tsetse – A fly endemic to sub-Saharan Africa best known for transmitting Trypanosomiasis ('sleeping sickness').

Xanthomonas – A genus of bacteria that produces a yellow pigment not soluble in water. Most xanthomonas species are pathogenic to plants.

Xylella fastidiosa – A type of bacteria that causes citrus variegated chlorosis (CVC).

List of Acronyms and Abbreviations

AAEA	American Agricultural Economics Association
AAPRESID	Argentine Association of Farmers for Direct Planting
ABS	Africa Biofortified Sorghum project
ABSP	Agricultural Biotechnology Support Project
AgBEE	Agricultural Black Economic Empowerment
AMS	Agricultural Marketing Service
ANVISA	National Agency of Sanitary Surveillance
APHIS	Animal and Plant Health Inspection Service
ARC	Agricultural Research Council
ARPOV	Argentine Association for the Protection of Plant Breeding
ARS	Agricultural Research Service
BASIC	Building African Scientific and Institutional Capacity
BC	National Agricultural GMO Biosafety Committee
BECA	Biosciences East and Central Africa
BMO	GMO Biosafety Management Office
BRIC	Biotechnology Regional Innovation Centre
Bt	*Bacillus thuringensis*
CAAS	Chinese Academy of Agricultural Sciences
CAFi	Chinese Academy of Fisheries
CAS	Chinese Academy of Sciences
CATA	Chinese Academy of Tropical Agriculture
CBMEG	Centre for Molecular Biology and Genetics
CBPP	contagious bovine pleuropneumonia
CCAP	Centre for Chinese Agricultural Policy
CEFOBI-Conicet Studies CENARGEN	The Center of Photosynthetic and Biochemical scientific and technological node of Brazilian agricultural biotechnology
CENIT	Centro de Investigaciones para la Transformación
CEPLAC	Executive Comission for Cocoa Crop Planning
CERAAS	Regional Centre for the Study of Crop Adaptation to Drought
CEVAN-Conicet	Center of Animal Virology at the Campomar Foundation

CGIAR	Consultative Group on International Agricultural Research
CGRFA	Commission on Genetic Resources for Food and Agriculture
CIARA	Chamber of Vegetable Oil Producers
CIRAD	Agricultural Research Centre of International Development
CITES	Convention on International Trade in Endangered Species of Wild Fauna and Flora
CNPq	National Research and Development Council
CONABIA	National Advisory Commission on Agricultural Biotechnology
CORAF-BBP	CORAF Biotechnology and Biosafety Programme
CORAF/WECARD	West and Central African Council for Agricultural Research and Development
CPB	Cartagena Protocol on Biosafety
CSIR	Council for Scientific and Industrial Research
CTNBio	National Biosafety Technical Committee
CUTS	Consumer Unity and Trust Society of India
CVC	citrus variegated chlorosis
D&PL	Delta and Pine Land
DACST	Department of Arts, Culture, Science and Technology
DBSA	Development Bank Southern Africa
DBT	Department of Biotechnology
DG	Directorate-General
DG Sanco	Department of public health and consumer protection
DPI	Brazilian Intellectual Property Rights System
DST	Department of Science and Technology
ECOPA	European Consensus – Platform for Alternatives
ECOWAS	Economic Community of West African States
EFSA	European Food Safety Agency
EIA	Environmental Impact Evaluation Study
EMBRAPA	Brazilian Corporation for Farming and Livestock Research
EP	European Parliament
EPA	Environmental Protection Agency
EPO	European Patent Office
ERS	Economic Research Service of the USDA
ESALq/USP	Agronomy School/University of São Paulo
ETC Group	Erosion, Technology and Conservation Group
EU	European Union
FABI	Forestry and Agricultural Biotechnology Institute
FANRPAN	Food, Agricultural and Natural Resource Policy Analysis Network
FAPESP	Foundation of Research Support of the State of São Paulo

FAO	Food and Agriculture Organization of the United Nations
FARA	Forum for Agricultural Research in Africa
FDA	Food and Drug Administration
FINEP	Project Finance Agency
FNSEA	Federation Nationale des Syndicats d'Exploitants Agricoles
GATT	General Agreement on Tariffs and Trade
GDP	gross domestic product
GEAC	Genetic Engineering Approval Committee
GLP	good laboratory practice
GM	genetically modified
GMO	genetically modified organism
GMOBC	GMO Biosafety Committee
GMV	genetically modified variety (of a certain crop)
GoI	Government of India
GTZ	German Technical Co-operation Agency
HSRC	South African Human Science Research Council
IBAMA	Agency for Environmental Regulation
IBC	institutional biosafety committee
IBGE	National Institute of Geography and Statistics
ICAR	Indian Council of Agricultural Research
IDEC	Consumers Defense Institute
IFPRI	International Food Policy Research Institute
IICA	Interamerican Institute for Agriculture Cooperation
IITA	International Institute of Tropical Agriculture
INASE	National Institute of Seeds
INERA	Institut de l'Environnement de Recherche Agricoles
INGEBI-Conicet	Institutes of Genetics and Biotechnology
INTA	National Institute of Agricultural Technology
IP	identity preservation
IPR	intellectual property rights
ISAAA	International Service for the Acquisition of Agri-biotech Applications
ISNAR	International Service for National Agricultural Research JGI
Research JGI	Joint Genome Institute
JMM	Joint Ministerial Meeting
JRC	Joint Research Centre
KZN	KwaZulu Natal
LPC	Law of Cultivars Protection
MAPA	Ministry of Agriculture, Cattle-Raising and Provisioning
MCT	Ministry of Science and Technology
MDG	Millennium Development Goal
MDP	Monsanto, Delta and Pine Land

MMA	Ministry of Environmental Affairs
MMB	Mahyco Monsanto Biotech
MOA	Ministry of Agriculture
MOH	Ministry of Health
MOST	Ministry of Science and Technology
MST	Landless Workers' Movement
NARS	national agricultural research services
NBN	National Bioinformatics Network
NGO	non-governmental organisation
NKL	national key laboratory
OECD	Organisation for Economic Co-operation and Development
ONSA	Organization for Nucleotide Sequencing and Analysis
ORF	Parapox viruses
PBS	Programme for Biosafety Systems
PlantBio	National Innovation Centre for Plant Biotechnology
PLRV	Potato Leaf Roll Virus
PPP	purchasing power parity
PRONEX	Excellency Research Centers Support Program
PRSV	Papaya Ring Spot Virus
PUB	Public Understanding of Biotechnology
PVC	plant variety certificate
PVP	Plant Variety Protection (Act)
PVY	Potato Virus Y
RABSAC	Regional Approach to Biosafety for Southern African Countries
R&D	research and development
RCGM	Review Committee on Genetic Manipulation
REDBIO	Technical Cooperation Network on Biotechnology
RET	Special Temporary Registry
SAASTSA	South African Agency of Science and Technology Advancement
SABS	South African Bureau of Standards
SADC	Southern Africa Development Community
SAES	state agricultural experiment station
SAFeAGE	South African Freeze Alliance on Genetic Engineering
SAGENE	South African Committee for Genetic Experimentation
SAGPyA	Secretariat for Agriculture, Livestock, Fisheries and Food
SCMV	strawberry marginal chlorosis virus
SECYT	Secretariat of Science, Technology and Productive Innovation
SENASA	National Agricultural Food Health and Quality Service

SNPA	National System for Research in Farming and Cattle-Raising
SOE	state-owned enterprise
SPS	Agreement on the Application of Sanitary and Phytosanitary Measures
SY	science years
TRIPS	Agreement on Trade Related Aspects of Intellectual Property Rights
TSWV	Tomato Spotted Wilt Virus
UC	University of California
UESC	State University of Santa Catarina
ULV	ultra low volume
UNCTAD	United Nations Conference on Trade and Development
UNDP	United Nations Development Programme
UNICAMP	University of Campinas
UPOV	Union for the Protection of New Varieties of Plants
USAID	United States Agency for International Development
USDA	United States Department of Agriculture
USFCAR	Federal University of São Carlos
USP	University of São Paulo
WCA	West and Central Africa
WHO	World Health Organization
WTO	World Trade Organization

Preface

Global policy debates about agricultural biotechnology have gone amiss. Focused on trade disputes, the rights of consumer choice, the rights of organic farmers in Europe and the US, and corporate marketing tactics vs rights of farmers, current policy debates miss out on the central role of technology in development as a source of increasing productivity that opens opportunities for people and for developing countries to become internationally competitive. The public mistrust of scientists and corporations has stigmatized genetically modified (GM) crops as a technology and has closed the doors to exploring an alternative vision of how this powerful new technology can be harnessed for development. And a narrow focus on poverty has missed the real threat for developing countries – that this will be a source of more global divides between rich and poor countries, and between developing countries that have advanced science and technology infrastructure and those that do not. Global divides in the 21st century will be driven not only by inequalities in resources and governance, but surely by access to knowledge and the capacity to manage technology.

This book is an attempt to stimulate attention towards an alternative vision for GM crops and development. It grew out of my earlier work on globalization and technology for the *Human Development Reports* 1999 and 2001. Reactions to the 2001 Report – *Making New Technologies Work for Human Development* – were surprising, not because of the strong critique from the leading anti-GM groups, but because these critiques did not address the report's analysis of risk and institutional challenges. Moreover, these did not align with reactions from donor governments and agencies. This led me to see the gap for alternative visions and alternative strategies.

I am grateful to all those from whom I have learned about this subject in the course of the last several years. The first on this list is Calestous Juma, whose contribution to the 2001 Report was eye opening, and whose deep knowledge of technology and development has been a source of continuous inspiration. I have also learned much from others on the 2001 Report team, the Millennium Project Task Force on Technology, colleagues at the Kennedy School, and many in civil society groups that have been campaigning for technology for development. They are too numerous to cite individually, but I would like to single out Kate Raworth, whose work on the 1999 and 2001 Reports helped me understand much of what I know today.

Work on this book was carried out during my tenure as fellow at the Kennedy School of Government, Harvard University during 2004–2006, hosted at the Belfer Center and the Center for International Development. I am grateful to Calestous Juma, Bill Clark and Nancy Dickson for their valuable guidance and support during that time and to Brian Torpy and Pat McLaughlin for administrative support. I am also grateful to the Rockefeller Foundation, whose grant supported this work, including the workshop held at Bellagio in May 2005, and to the United Nations Development Programme (UNDP) for granting me sabbatical leave during the first fellowship year. I benefited from interviews with many who are engaged in policy debates from diverse points of view – non-governmental organizations (NGOs) in Europe and the US, government agencies and politicians in Europe, the World Bank and UNDP, Consultative Group on International Agricultural Research (CGIAR) scientists, and academics. Robert Paarlberg, Diane Osgood, Carl Pray, Debbie Delmer and Smita Srinivas gave me valuable comments. Mary Lynn Hanley was a stellar editor. Nina Dudnik and Ben Roberts provided timely and reliable research support. Finally, I am grateful to the contributing authors to this volume for participating in what developed into a collective effort based on professional collegiality and personal regard.

Thanks also to Francis, my husband, and to my sons Nicholas and Henry for being there.

Sakiko Fukuda-Parr
Cambridge, Massachusetts
June, 2006

Part One

National Development Priorities and the Role of Institutions: Framing the Issues

Introduction:
Genetically Modified Crops and
Development Priorities

Sakiko Fukuda-Parr

Why this book?

Investing in agricultural technology increasingly turns up these days on lists of the top ten practical actions the rich world could take to contribute to reducing global poverty. This is surprising because international development support to agricultural research has been declining over the 1980s and 1990s. Why this new attention? Investing in technology for tropical agriculture, as for tropical health, creates global public goods that neither markets nor developing country governments can supply on their own, and requires global science and finance. Besides, agriculture is a source of income for the world's poorest – supporting about 70 per cent of those living on less than US$1 a day. These farmers have not benefited from the technological innovations of the last half century. Their per hectare yields of major food crops have not increased in Africa over the last three decades even as they have doubled or quintupled in other regions. Modern science and technology can surely be applied to solve problems that keep their productivity low. While agricultural technology is location specific, farmers and national agricultural research efforts need not innovate on their own. The Green Revolution has shown how high-yielding varieties developed at international centres can be adapted to local conditions, dramatically increasing yields and farm incomes. Past investments in agricultural technology have a documented record of extremely high economic returns. Economists have argued that there is underinvestment in agricultural research and development; spending from both national and donor sources has been declining over the last two decades in most countries in Africa, Latin America and Asia, except for China, Brazil and a handful of other countries that have emphasized science and technology as a strategic pillar of development (Beintema and Pardey, 2001).

All these points make a compelling case for reviving global research efforts to develop technology that will improve the productivity of small-scale, resource poor farmers. In addition, science adds another factor: breakthroughs in biotechnology have opened up enormous new possibilities. Applications in agriculture make it possible to develop more efficiently new crop varieties with tolerance to saline soils, drought conditions or pest attacks, thus addressing constraints that have dogged farmers for centuries.

Yet this technology is not a visible part of the international development agenda because there is too much controversy about its risks and relevance for developing countries and their traditional farming sectors. Advocates herald the promise – as yet unfulfilled – of new varieties that could increase farm productivity and human nutrition. Opponents warn of threats of farmers losing control of their livelihoods as they give in to corporate takeover of agriculture that irreversibly damages the ecosystem. Each side claims the moral high ground and argues on behalf of the interests of poor people in poor countries, but their arguments focus on predictions of what might happen in the future. How can this technology with so much potential be governed to work for development – for growth, equity and sustainability?

In addressing this important question, we should start by gaining a better understanding of the experience of developing countries to date. This is the subject of this volume, which documents the experience of five countries that lead commercial production of genetically modified (GM) crops, namely Argentina, Brazil, China, India and South Africa. Behind the international controversies, and less in the public eye, is the reality that developing countries have already begun to adopt the new technology. Over 60 countries are conducting research programmes with agricultural biotechnology applications (Cohen, 2005) with substantial scientific and technological mastery, and a handful have started commercial production of GM crops.

The aim is to compare these experiences in order to understand the process of this technological diffusion and draw lessons from it, in particular how these countries have used the technology to pursue their national development priorities and especially the role of institutions and government policies in this process.

What are genetically modified crops?[1]

GM crops and varieties, also referred to as '*transgenic crops*' and '*genetically modified organisms (GMOs)*', are developed by a process of genetic modification that allows selected individual genes to be inserted from one organism into another to enhance desirable characteristics ('traits') or to suppress undesirable ones.

The process of genetic modification is also called '*genetic engineering*', '*gene technology*' or '*recombinant DNA technology*'. It may involve genes from related or unrelated species. Genomics refers to the comprehensive study of the interactions and functional dynamics of whole sets of genes and their products.

Genetic modification is only one of the many techniques or tools of '*modern plant biotechnology*'. Two other tools are commonly applied. The first is *tissue culture*, in which new plants are grown from individual cells or the culture of cells, often bypassing traditional cross-fertilization and seed production. The second is *marker aided selection*, in which DNA segments are used to mark the presence of useful genes, which can then be transferred to future generations through traditional plant breeding using the markers to follow inheritance. The use of these two other techniques for crop improvement is much less controversial than genetic modification. This book, as the title indicates, is about GM crops rather than agricultural technology more broadly. It is a complement to – not a substitute for – many areas of conventional agricultural research.

While the novelty of transferring a gene from one plant (or even an animal) to another is a source of suspicion and leads to fear of GM crops, scientists emphasize that biotechnology is centuries old. From the time that human crop cultivation began, farmers have attempted to improve the genetic makeup of plants to enhance their productivity, especially their yields, their tolerance of droughts and other characteristics. The history of these endeavours evolved over the centuries through three stages. First came the collection of landraces; farmers and communities developed well-performing varieties through *selection*. The second phase started in the 1900s with Mendel's discovery of genetic principles. With the scientific application of Mendelian principles to *conventional breeding* involving cross breeding of varieties, more targeted results could be obtained. The third phase, '*modern biotechnology*', uses the tools of biotechnology.

The key difference between a GM crop variety and an improved variety created by conventional plant breeding is that in a GM organism, the genes are transferred without sexual crossing. In addition, in conventional breeding, the specific genes controlling a trait may not be identified, whereas in genetic engineering, well-characterized genes are being transferred in a targeted manner. One of the most fundamental and contested questions is whether these organisms are substantially different from non-genetically engineered varieties and pose greater risks to human health and environmental sustainability. There is voluminous literature on the diverse scientific evidence and arguments on these issues. But the need for effective stewardship is well recognized by all – even by industry lobbyists – and countries have passed legislation to regulate research and commercial release based on criteria related to impact on human, environmental and agricultural practice.

The key advantage of genetic modification is that it makes the process of crop improvement more efficient, taking less time and using more precise processes. Applying genetic principles has enabled scientists to cut down the years needed to develop desirable traits or eliminate undesirable ones by selection. But generations of cross breeding are still necessary to find the desirable traits.

Genetic modification allows scientists to manipulate much more precisely the genetic material, adding precision and dramatically reducing the time needed for development work. Because one can transfer genes across species, something not possible in conventional breeding, genetic modification also vastly expands the scope of what is possible in developing a variety with desirable characteristics, and thus expands the frontiers of research.

Investments in research and development (R&D) have led to development of commercial varieties of soy, cotton, maize and canola carrying herbicide resistant or insect resistant genes. These varieties were initially developed in the US and targeted for large-scale American agriculture, but they have now spread across the world and are under commercial production in some 21 countries. These developments are described in Chapter 2.

Developing country priorities

Five policy objectives: Beyond poverty and hunger

World hunger and poverty drive the polemics concerning the potential benefits and risks of biotechnology for developing countries. While advocates emphasize the potential for increasing production, opponents emphasize the threats to rural livelihoods from potential risks of ecosystem changes and dependence on corporate control of the seed supply and the spread of industrial agriculture.

GM technology can increase farm productivity and could possibly reduce poverty and hunger. But there are no automatic links, and the impact of introducing GM crops would depend on a number of factors. Not all GM crops will increase production; herbicide tolerant varieties such as RR soy would not necessarily lead to production increases but would reduce costs.

Neither will all agricultural technology be beneficial for resource poor small-scale farmers. As lessons of the Green Revolution have shown, its impact depends on three conditions: first, that the technology is relevant to those farmers and affordable; second, that the technology is scale neutral; and third, that the existing socio-economic environment (such as access to markets, information or inputs) is not heavily biased against small-scale farms (Hazell and Ramasamy, 1991). So far, the GM crops developed and released for commercial production were developed for US farmers and involved the crops and traits most relevant to their production constraints. For resource poor and small-scale farmers in developing countries, the key food crops would include rice, wheat, maize, cassava, plantain, millet, sorghum and legumes, while constraints would include low yield, drought, pests, diseases and low soil fertility.

So far, the GM crops developed and put under commercial production are not – with the exception of maize – the major staple food crops of the most marginal farmers in developing countries (e.g. cassava, banana, sorghum and millet). Many advocates of agricultural biotechnology emphasize hunger reduction as a key motive for putting this technology to use in developing countries. Not only Monsanto, the multinational company, but Norman Borlaug, the scientist, and Per Pinstrup Anderson, the international authority on food and hunger, draw a parallel with the Green Revolution to argue that food production needs to increase to keep up with population growth. However, as noted by Ramaswami in Chapter 9, this is predicated on a Malthusian analysis of hunger. Many argue that hunger results not from an insufficient volume of food production, but because poor people do not have access to it owing to lack of income, or to other social provisions to ensure them food. Globally, some 852 million people are hungry (UN Millennium Project, 2006), not because there is not enough food produced in the world but because they do not have the income to purchase it.

While growing GM crops developed to date can reduce use of chemicals, these are not necessarily the most relevant environmental risks faced by the most vulnerable people in developing countries. Desertification, soil erosion and salinization are some of their major problems, while risks of ecosystem changes caused by growing GM crops have been highlighted as major issues.

Thus the links between adoption of GM technology and development depend on what crops are developed, with what traits and for which farmers. In short, it is R&D and marketing priorities that shape those results; while market incentives would drive these priorities in the private sector, it is government policy that can influence the path of technology development.

But developing countries have objectives beyond poverty. They have several different reasons for promoting technological innovation that improves agricultural productivity as part of their national economic growth and development strategies:

1   To increase aggregate production and accelerate growth of the gross domestic product (GDP) and employment. In most developing countries agriculture plays a major role in the economy, contributing to exports, employment and farm incomes. For example, in 2004 it accounted for 16 per cent of the GDP in sub-Saharan Africa, 21 per cent in South Asia, 15 per cent in China and 22 per cent in India, in contrast to just 2 per cent for the high income Organisation for Economic Co-operation and Development (OECD) countries (World Bank, 2006). Moreover, agricultural productivity is low in most countries. Cotton yields, for example, are lower in India than almost anywhere else in the world.

2   To increase farm household incomes and reduce poverty. In most developing countries, the poorest people are often those living in rural households. Increasing productivity improves farm incomes for poor rural households and, if the employment effect is positive, for landless labourers. Globally, it is estimated that some 70 per cent of the 1.2 billion people living on less than US$1 a day are in rural areas living off agriculture (UN Millennium Project, 2006). Empirical evidence has also shown the importance of increasing agricultural

productivity and agricultural GDP in generating growth that has higher poverty reducing effects. The evidence presented in Chapter 9 is an example of the evidence documented for many other countries (UNDP, 1997).

3   To increase food production, reduce hunger and improve food security. Many of the 852 million or more hungry people in the world today live in farm households or low income urban households where more than half of the household budget is spent on food. Improved technology can contribute to reducing hunger by increasing food supplies for subsistence farmers, improving incomes of farmers and reducing prices for urban consumers.

4   To reduce use of chemical inputs and improve human health and environmental sustainability. Chemical inputs raise yields but are a risk to health and the environment. Chemicals are often handled without safeguards and farmers risk poisoning.

5   To participate in the forefront of scientific and technological progress and compete in global markets. Though not all countries will be leading science and innovation, all need the capacity to innovate and to be able to access and use global technology that is useful in meeting their national needs. As Hayami and Ruttan have argued, success in achieving a dynamic agricultural sector depends on 'the capacity to generate an ecologically adapted and economically viable agricultural technology in each country or development region' (Hayami and Ruttan, 1986). This requires a dynamic process of continued adjustment to changing economic and social conditions. Being able to access global technology and having the capacity to be technologically competitive is an important part of being integrated into the global economy. Some developing countries have taken a policy decision to position themselves as leaders in technology alongside global innovators such as the US and Japan.

Agenda for global integration and social equity

The above five objectives are all important aspects of challenges that developing countries face. The first and last objectives (increasing production and participating in cutting-edge science and technology) are important parts of a global integration agenda. In the context of globalization, an important priority for developing countries is the ability to compete in world markets; accessing technology is a critical means to participate in, rather than become marginalized from, the global economy. The other objectives are important parts of a social equity and sustainability agenda. But all of them are important means towards achieving the ultimate purpose of development, namely expanding human well being in its broad sense of the ability of individuals to lead full lives, or human development in the full sense of its meaning. Access to productivity enhancing technology can expand human lives by increasing incomes, which opens up many other opportunities.

Among the five objectives, those relevant to a country depend on the particular conditions of that country, which can differ considerably:

- A country where soy, cotton, maize and canola are economically and socially important crops could use available GM crops if the traits addressed production constraints.
- A country wishing to pursue GM crop technology for poverty and hunger reduction would need to develop GM varieties of food crops and main cash crops and breed them for yield, drought resistance and pest resistance.
- A country wishing to pursue GM crops for major exports of cotton, soy, canola or maize would have to compete with GM crop exporters in global markets and must therefore consider the risks of *not* adopting the technology. Higher productivity of GM crops would lead to increased production by the US and pressure on prices that may squeeze out the country's exports.
- A country strategically placing itself in the forefront of science and technology could not afford not to have research programmes in agricultural biotechnology.

Concerns of opposition movements

Opposition groups have pressed for outright bans or moratoria on GM crops.[2] The many issues raised can be grouped into three areas of risk:

1   ecological and human health risks;
2   socio-economic risks;
3   cultural risks of foreclosing consumer and public choice.

First, ecosystem changes and harm to human health; these risks have been at the heart of the controversies about agricultural biotechnology from the very beginning (Osgood, 2001). Controversies over these concerns drive political struggles over legislation on procedures for approval of GM crops, with the opposition groups pushing for restrictive legislation based on 'precautionary' principles. Over the years, evidence has accumulated, public debates have advanced and a consensus has emerged among the scientific establishment, including institutions such as the American Academy of Science, the Third World Academy of Science, the Royal Society, the Nuffield Council of Bioethics and the Food and Agriculture Organization of the UN (FAO). This consensus is that the crops released for commercial production so far are safe for human consumption; that environmental risks should be assessed on a case by case basis; and that over the last decade of commercial production, no cases of ecosystem consequences have been experienced (FAO, 2005).

Opposition groups argue that these conclusions are not based on sufficient analysis and that the technology has been prematurely released (Roberts, 2006). Driving this position is a fundamental mistrust of the official scientific establishment; sometimes allegations are made that it has been influenced by industry lobbies.[3] Many commentators observe that mistrust of government scientists and regulators is particularly strong in Europe, with its recent history of scientific judgements proving wrong, such as the outbreak of mad cow disease in the UK after it was initially denied by government scientists. But such cases have also occurred elsewhere, such as in the US, which has a history of pharmaceutical products being withdrawn from the market after passing Food and Drug Administration (FDA) approval; or in Japan, which experienced scandals of foods being mislabelled.

Second, socio-economic risks from corporate control of GM crop variety and seed supply; GM crop technology is hard to disentangle from corporate power, which has been behind its development and diffusion. To quote one anti-globalization group, 'Too much power and control over the world's agriculture and food system is ending up in the hands of too few and purely commercial interests' (Christian Aid, 2005). While the potential of genetic engineering in developing GM crops to benefit poor farmers in developing countries is not denied, opposition movements argue that the corporate control of R&D and its influence on national biosafety legislation, global trade and biosafety agreements ultimately puts in place a process of agricultural transformation, from traditional to commercial agribusiness type agriculture (ETC Group, 2006; Grain, 2006). This is a true threat to the survival of traditional livelihood systems that could work through a variety of channels. For example, GM crops could pose a threat to the ecosystem by making changes that could undermine biodiversity, important for indigenous communities and their livelihoods. Farmers could become dependent on commercial seeds rather than being able to save their own, thus losing the ability to manage their farming systems and having no means to control new problems that might arise, such as pest infestations. Family farms could be squeezed out due to economic competition from the more productive agribusiness farms. Yet these risks could just as well apply to the introduction of high yielding varieties produced through conventional breeding; the real difference with the GM varieties is the corporate profit that drives their development and diffusion and the political economy of the process.

Most research-based empirical studies of socio-economic impact have focussed on *short-term farm level* benefits (e.g. income, production (yield), resistance to insects, health of farmers). Most such analyses show GM varieties to be distributionally neutral or to deliver more benefits to smaller scale farms than larger ones (FAO, 2005). Studies also show that the welfare gains from GM crops are not monopolized by multinational seed companies but shared by the farmers (FAO, 2005). But these studies do not respond to the concerns raised by the anti-GM groups that are about longer term, unintended and secondary impacts on the environment and social structures.

Third, cultural risk of closing consumer choice; for many people, GM crops are unnatural and are rejected as a matter of values and culture. They argue that it is a matter of values to restrict the application of science to such unacceptable ends, and that they have a right to exercise choice over the foods they consume. This opposition, a matter of defending social norms or 'culture', is expressed by the public in many countries around the world. It is particularly important politically in Europe, where policy debates focus on efforts to protect a GM-free market through mechanisms such as labelling of products and requirements for distance between GM, conventional and organic agriculture. The rejection of GM crops on these value grounds builds on movements that have advocated turning the tide of chemical-intensive, agribusiness-dominated agriculture, protecting family operations, and developing alternative farming methods – organic and otherwise – that depend less on chemical applications. These concerns relate to issues of science and society and are about decision-making processes that shape technological trajectories, which contemporary sociologists and other theorists such as Jasanoff have begun to address (Jasanoff, 2005).

These three areas of concern are advocated by national and international NGO networks. These include 'environmental groups' such as Friends of the Earth and Greenpeace, 'anti-globalization' development groups such as the Erosion, Technology and Conservation (ETC) Group, Third World Network, Christian Aid, and consumer groups such as Consumer Unity and Trust Society of India (CUTS) (Osgood, 2001). While the concerns can be separated into three areas, they are seen to be interconnected because they are ultimately tied up with the political economy of this technology – the fact that it is spread through a process driven by corporations motivated by corporate profit. Thus multinational seed companies not only invest their massive resources to develop the technology, they appropriate it using strong intellectual property laws, and lobby for non-restrictive or 'permissive' biosafety legislation at national and global levels. They also campaign against labelling and influence World Trade Organization (WTO) trade negotiations (Wright and Pardey, 2006). By virtue of their size, corporations wield financial might. They have grown even larger as the sector has become increasingly consolidated with mergers and acquisitions. There are now only seven mega corporations operating in this market.

Local priorities, local concerns, local processes

These concerns highlight the critical role of national capacity – through its institutions and policies – to shape the use of this technology to meet national objectives. The strategy of the GM opponents is to shape legislation to restrict the supply and demand for GM varieties by pressing for restrictive biosafety standards, labelling and identity preservation. The central purpose of this book is to ask whether and how national institutions and policies can shape the use of this technology to serve national objectives, and the priority constraints facing farmers in developing countries.

The potential role of GM crops – both benefits and risks, present and potential – depends on the particular conditions of the country growing them, whether it can gain from increasing farm productivity, whether it has a science/technology advantage, and whether there are populations (such as organic farmers or indigenous people) who are vulnerable to ecosystem changes. In a country where soy, cotton, canola and maize are major crops and where herbicide resistance and insect tolerance can be of help because of its agro-ecological conditions and factor endowments, farmers growing those crops would benefit from GM technology. It would improve farm incomes for those farmers and increase exports and the agricultural GDP. Consumers might benefit from lower prices. But in countries where agriculture is not an important part of the GDP and increasing its productivity is not a priority, there is little benefit from this technology. As shown in the chapter on Europe, most European countries have opposed this technology. This is not surprising because European consumers stand to gain little if anything from it (except possibly lower prices) and feel uncertain about the science. For many, too, the idea goes against their vision of science's role in society. In such countries the main sector of society that benefits is the farmer growing soy, cotton, maize or canola who has not opted for organic methods – and these make up a small proportion of the country's population. Scientists also benefit, as do multinationals involved with GM crops. This explains the ambivalence of some countries – particularly Switzerland, home of the Syngenta corporation.

The plan of the book

The case studies of five developing countries – Argentina, Brazil, China, India, South Africa – are at the heart of this book. They are contained in Part II (Chapters 5–10). The book starts with chapters that set up the background, concepts and issues that frame the analysis. So Part I (Chapters 1–4) contains this introduction and a chapter on the emergence and diffusion of GM technology at the global level focusing on institutions and markets as key drivers. Part I also contains three case studies presenting sharply contrasting experiences that illustrate the role of markets and institutions – the US, which continues to lead the technological development, the European Union (EU), which invested initially then diverged, and West and Central Africa (WCA), which has developed a plan that has yet to be implemented. Part III (Chapters 11–12) of the volume compares and analyses the institutional change and role of policy in the five countries. The final chapter concludes the volume by revisiting the question of government policies and prospects for this technology to follow a pro-poor path.

Each of the case studies is written by a leading national researcher working out of a national research institution. The goal is to reflect the perspective of nationally defined development priorities, and to assess the role of GM technology in the context of meeting them rather than in relation to a pre-defined set of priorities.

Unfortunately, no study on the theme of GM crops can escape from the polemics that have polarized the debate and this volume will inevitably be read as supporting one camp or the other. Opposition groups will see the book as pro-GM from the very fact that the study is not advocating stopping the further development of the technology. Alternatively, advocates of the technology will regard it as inadequate because it raises too many questions about GM crops' potential impact on poverty reduction and corporate ownership and does not advocate its spread. The purpose here is not to reinforce one or the other of the two extreme positions, but to contribute to developing a middle ground: an understanding of the institutional and policy innovations that can make this technology work for development.

Each of the contributing authors has a particular position on the numerous issues in the debate; being a part of this book does not mean that these authors have a common position or are in agreement. What they share is a desire to contribute to development and to the well being of people in developing countries, and a serious interest in exploring the experience of the five countries to date to identify how – and if – GM crop varieties can be a technology for development.

The volume does not focus on evaluating the impact of GM crops on development because the experience has been too limited to make such an assessment. While there have been numerous studies of impact on farm incomes, pesticide use and farmer health, these have covered only a few years. The issues that drive this controversy are the health, environmental and socio-economic risks of foods that may be toxic or cause allergies, environmental consequences on reducing biodiversity, and the socio-economic consequences on livelihood systems, including dependence on purchase of seeds, as well as secondary impacts that technological innovations bring such as shifts in employment and land ownership patterns. These impacts are systemic and take a longer time to materialize. The purpose of this book is to reflect on the empirical evidence and learn lessons for how a pro-development, pro-poor path could be forged for agricultural biotechnology.

Notes

1   See FAO (2005) for a more detailed explanations of the science and technology of agricultural biotechnology and genetic modification, drafted in accessible language for non-specialists.

2   This section is based on an extensive review of websites of opposition groups including Consumer Society International, Friends of the Earth, Greenpeace, ETC Group, Grain and others. This review is summarized in Roberts (2006). See Osgood (2001) for an overall analysis of the civil society opposition to GM crops. Although this analysis focuses on the situation up to 2000 and the situation has evolved, it has not been radically altered.

3   See, for example, the letter mobilizing the moratorium campaign of April 2005 (Christian Aid, 2004).

References

Beintema, N. and Pardey, P. (2001) *Slow Magic, Agricultural R & D a Century After Mendel*, International Food Policy Research Institute, Washington DC

Christian Aid (2005) 'Christian Aid and the GM crops debate', 20 December, available at www.christian-aid.org.uk/indepth/412gmfood/index.htm, accessed 20 March 2006

Cohen, J. (2005) 'Poor nations turn to publicly developed GM crops', *Nature Biotechnology*, vol 23 (1), pp27–33

FAO (2005) *State of Food and Agriculture 2005 – Agricultural Biotechnology: Meeting the Needs of the Poor?* Food and Agriculture Organization, Rome

Hayami, Y. and Ruttan, V. (1986) *Agricultural Development, An International Perspective*, Johns Hopkins Press, Baltimore

Hazell, P. and Ramasamy, C. (1991) *The Green Revolution Reconsidered: The Impact of High-Yielding Rice Varieties in South India*, The Johns Hopkins University Press for IFPRI, Baltimore

Jasanoff, S. (2005) *Designs on Nature, Science and Democracy in Europe and the United States*, Princeton University Press, Princeton

Osgood, D. (2001) 'Dig it up: Global Civil Society's responses to the plant biotechnology agenda', in Anheir, H., Glasius, M. and Kaldor, M. (eds) *Global Civil Society 2001*, Oxford University Press, Oxford

Roberts, B. (2006) 'Summary of opposition arguments to GMOS', Working note

UN Millennium Project (2005) *Halving Hunger: It can be done.* Report of the Task Force on Hunger, Oxford University Press, Oxford

UNDP (1997) *Human Development Report*, Oxford University Press, New York

World Bank (2006) *World Bank World Development Indicators*, available at http://devdata.worldbank.org/external/CPProfile.asp?PTYPE=CP&CCODE=ZAF

Wright, B. and Pardey, P. (2006) 'The evolving rights to intellectual property protection in the agricultural biosciences', *International Journal of Technology and Globalisation*, vol 2 (1/2), pp12–29

Emergence and Global Spread of GM Crops: Explaining the Role of Institutional Change

Sakiko Fukuda-Parr

Biotechnology enthusiasts emphasize the power of the new science to address a seemingly endless array of constraints facing resource poor farmers. But the process of technological innovation depends as much on institutions as on the science (Hayami and Ruttan, 1985). Recent theories of technological change increasingly focus attention on the 'co-evolution' of technology and institutions and the role of public policy in that process (Nelson, 1994; Mokyr, 2002). GM crops present new challenges of institutional innovation as much as scientific innovation. They were developed with a new breakthrough science, but also with a radically new institutional model. In a sharp departure from the history of public sector-led research and development, the 'gene revolution' is led by a handful of mega-corporations.

How can developing countries manage the use of this powerful technology, driven by powerful corporations, for their own priorities? Does the new institutional model for GM technology necessarily have to follow the identical US corporate model? What is the environment that will provide the incentives for investing in priority crops and traits? Understanding the role of institutions in GM crop development would be a first step to answering these questions. This chapter analyses how institutional responses shaped the global pattern of development and diffusion of GM crop technology over the last decade. The chapter draws on the three case studies in this section – US, Europe and West/Central Africa – to illustrate this process. The divergent technological trajectories of these cases came with sharply contrasting institutional responses.

History of GM crops: Emergence and global spread

Research and development

The technology of genetic modification was developed in the 1980s following the Nobel Prize winning work of scientists Cohen and Boyer on recombinant DNA and other breakthroughs. The 1980s saw a surge of private investments in applications of this technology for both health and agriculture, first by start-up companies financed by venture capital, and then by multinational corporations. In agriculture, agro-chemical corporations such as Monsanto and Syntenta began to invest massively in developing GM varieties. These companies had no scientific expertise in either biotechnology or plant breeding. They invested in acquiring these capacities by recruiting new staff and by merging with or acquiring seed companies that had experience and know-how in plant breeding and the commercialization of seeds. Box 2.1 describes the scientific processes that are required and how the multinational agro-chemical companies managed to combine their financial and scientific resources with the genetic resources of the seed companies.

Box 2.1 The scientific process for generating GM crop varieties

Greg Traxler

A handful of vertically coordinated firms have been the key players in ushering in the biotechnology revolution in the US and other countries. These firms have been successful in linking useful genetic events with high quality germplasm to create GM varieties with the ability to gain rapid market penetration and to capture value for their creators. These firms have used mergers, acquisitions and licensing agreements to ally their financial, scientific and organizational strengths with the genetic resources of traditional seed companies such as Delta and Pine Land, Asgrow, Pioneer, Dekalb and dozens of smaller ones.

The creation of commercially viable GM varieties can be thought of as resulting from combining the products of two largely distinct scientific undertakings: a biotechnology step and a plant breeding step (see Figure 2.1). The biotechnology step produces a genetic event or gene transformation that is useful in solving an economically important agricultural problem. The gene must then be combined with an adapted crop variety to create a viable commercial GM variety. The two steps are largely separate scientific enterprises and need not occur in the same institution, or even in the same country. Most GM varieties today are marketed and delivered to farmers by seed companies that do not have the capacity to do genetic transformation. The seed companies license transgenes from Monsanto, or another biotechnology company, and use conventional breeding techniques to transfer the genes into their best commercial lines. Delta and Pine Land Seed Company, which is one of the world's leading marketers of GM varieties, has never had significant biotechnology research capacity. It is a modest-size corporation with eight US-based cotton breeders and three breeders in other countries.* However, the seed companies have extensive experience in marketing seeds to farmers – an important capacity that the multinational gene discoverers do not have.

The type of scientific capacity required to mount a successful biotechnology research programme is fundamentally different from that needed for developing adapted crop varieties. Discovering useful genes, transferring them to the intended plant species and achieving an adequate level of expression of the alien gene in the new plant host makes use of 'new' biotechnology science techniques, while the plant breeding step relies primarily on proven conventional plant breeding techniques. The science of biotechnology itself is evolving rapidly, with a steady stream of process innovations, so this distinction is becoming less valid and the line between biotechnology and plant breeding is becoming ever more blurred. For example, marker assisted selection is a technique developed by biotechnology research firms, but which is now routinely used by seed companies. The companies that have successfully commercialized transgenes have needed large investments, cutting-edge scientific talent and the skilled legal council required to negotiate intellectual property hurdles.

Success in discovering a marketable event is also very uncertain, with many more failures than successes. The end product of the biotechnology step is a crop variety with an adequate level of expression. The transfer of the gene to the initial, or receptor, variety is accomplished using one of several biotechnology protocols. Subsequent transfers to other varieties of the same crop are done using traditional plant breeding techniques.

Because the receptor variety is chosen on the basis of its characteristics in expressing the gene, rather than its superior agronomic performance, the plant breeding step is essential to achieve a GM variety that will be successful in the market. This step for self-pollinating crop varieties is straightforward. The receptor variety is crossed with a leading commercial line, followed by three or four backcross generations and another three selfing generations to attain a genetically stable variety. Using modern breeding practices, it is possible to produce several selection generations per year, so marketable GM varieties can be produced from a receptor line in 2–3 years. This plant breeding step of creating commercial GM varieties from a receptor GM variety is quicker, easier and more certain than the development of a commercially successful conventional variety from a pool of elite lines, and can be done at a fraction of the cost.

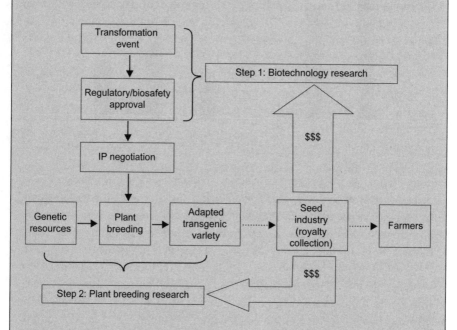

Figure 2.1 *The two-step scientific process for producing GMOs*

Note: *The totals include breeders working in Paymaster and Suregrow divisions. International breeders are located in Australia, Greece and Turkey.

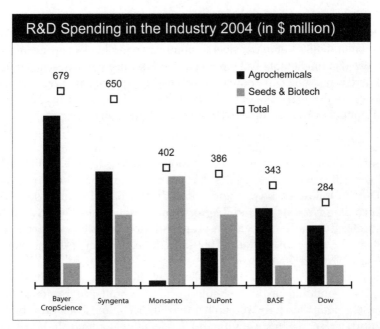

Figure 2.2 *R&D budgets of leading agricultural biotech multinationals (million Euro)*

Note: Spending on agricultural biotechnology only. Corresponding figures for total R&D spending in 2004 in US$ millions were as follows:

Company	R&D 2004
Syngenta	$809m
Bayer CropScience	$845m
Monsanto	$509m
DuPont	$1333m
Dow Chemical	$1022m
Delta & Pineland	$18.4m

DuPont acquired Pioneer.

Rhone-Poulenc Agro and AgrEvo merged in 2000 to form Aventis CropScience.

Bayer acquired Aventis in 2002, creating Bayer CropScience.

Ciba-Geigy and Sandoz merged into Novartis in 1996. Novartis Seeds merged with Astra Zeneca's agribusiness section in 2000 to form Syngenta.

Monsanto acquired Seminis as a subsidiary in 2005. Seminis had earlier acquired Asgrow, Petoseed, Royal Sluis, Horticeres.

Source: Company Report, Cropnosis

With the exception of *Bt* cotton, developed by the Chinese Agricultural University,[1] all of the major crops commercialized and under cultivation extensively to date[2] were developed by multinational chemical and seed companies. The only other GM crops that have been developed by the public sector are the disease resistant papaya variety developed by the US Department of Agriculture in Hawaii, and chilli, sweet pepper and tomato, developed in China, all of which are grown locally in limited quantities. Byerlee and Fischer (2001) estimated total global expenditures in agricultural biotechnology for 2000 in the range of US$2 billion to US$3 billion, showing high concentration in industrialized countries (about 90 per cent) and in the chemical/seed companies. The current pipeline of products is also highly concentrated in industrialized countries; of the 11,105 field trials conducted on 81 crops from 1987 to 2000, only 15 per cent were in the developing and transition countries, where public sector institutions are major players (Pray and Naseem, 2003). The world's top ten bioscience corporations spend nearly US$3 billion annually on research. Monsanto, the largest investor in crop biotechnology, spent US$588 million in fiscal year 2005 while its net sales that year totalled US$6.3 billion. Syngenta and DuPont follow, each spending over half of what Monsanto does. But the rest, including Dow, Bayer and BAgro, spend a lot less (see Figure 2.1). Among industrialized countries, US and European-based biotech companies invested heavily in the 1980s and into the 1990s but these investments fell sharply in the 1990s (see Chapter 4).

However, the picture is fast evolving as the public sector in developing countries is building up biotechnology capacity and programmes. Over 60 developing countries have research programmes, almost all in public sector agricultural research institutions and universities (Cohen, 2005). The most sizeable programmes are in China, Egypt, South Africa, Brazil and India (Pray and Naseem, 2003), with China's by far the largest. Huang et al estimate that the Chinese government's spending on agricultural biotechnology reached RMB1.647 billion in 2004, or US$199 million, equivalent to US$953 million in PPP$ (see Chapter 8). The annual biotechnology budget in Brazil, probably the next largest, is about US$30 million (see Chapter 7). One of the key questions that will be addressed is whether the public sector institutions in developing countries will become major centres of research and develop products most useful for developing country farmers.

Commercial production

The first GM crops were released and planted in 1994 in Canada, and then in the following year in the US and several other countries. But it was in 1996, in the US, that significant areas (1.45 million ha) were sown to GM soy, maize and cotton. Production in the 1990s was concentrated in the US, where it still accounts for about half of total global area. But from the initial years, over a million hectares were sown in Argentina, Canada and China (James, 2005).

Since then, the global areas under production have expanded rapidly, at over 10 per cent annually, reaching 90 million hectares in a decade, as shown in Tables 2.1, 2.2 and 2.3. By 2005, 21 countries were producing GM crops, including the US and Canada, seven countries in Latin America (Argentina, Brazil, Paraguay, Uruguay, Mexico, Colombia and Honduras), four in Asia (China, India, Iran and the Philippines), four in Western Europe (Spain, Portugal, Germany and France), two in Eastern Europe (Romania and the Czech Republic), one in Africa (South Africa), and Australia.

Table 2.1 *Commercial production of GM crops worldwide, 1996–2005: Total areas by country (million ha)*

Country	1996	1997	1998	1999	2000	2001	2002	2003	2004	2005	% of total
USA (soy, maize, cotton, canola, squash, papaya)	1.5	7.2	20.8	28.6	30.3	35.7	39.0	42.8	47.6	49.8	55
Argentina (soy, maize, cotton)	<0.1	1.5	3.5	5.8	10.0	11.8	13.5	13.9	16.2	17.1	19
Brazil (soy)	0	0	0	1.2	0	0	0	3.0	5.0	9.4	10
Canada (canola, maize, soy)	0.1	1.7	2.8	4.0	3.0	3.2	3.5	4.4	5.4	5.8	6
China (cotton)	1.0	1.0	1.1	1.3	0.5	1.5	2.1	2.8	3.7	3.3	4
Paraguay (soy)	0	0	0	0	0	0	0	0	1.2	1.8	2
India (cotton)	0	0	0	0	0	0	<0.1	0.1	0.5	1.3	1
South Africa (maize, soy, cotton)	0	0	0.1	0.2	0.2	0.2	0.3	0.4	0.5	0.5	0.6
Uruguay (soy, maize)	0	0	0	0	<0.1	<0.1	<0.1	<0.1	0.3	0.3	0.3
Australia (cotton)	0	0.2	0.3	0.3	0.2	0.2	0.1	0.1	0.2	0.3	0.3%
Mexico (cotton, soy)	0	0	0.1	0.1	<0.1	<0.1	<0.1	<0.1	0.1	0.1	0.1%
Romania (soy)	0	0	0	0	<0.1	<0.1	<0.1	<0.1	0.1	0.1	0.1%
Philippines (maize)	0	0	0	0	0	0	0	<0.1	0.1	0.1	0.1%
Spain (maize)	0	0	0	<0.1	<0.1	<0.1	<0.1	<0.1	0.1	0.1	0.1%
*											
Total	2.57	11.51	28.62	41.49	44.2	52.6	58.7	67.7	81.0	90.0	

Note: *Less than 50,000 ha grown in Colombia, Iran, Honduras, Portugal, Germany, France and Czech Republic in 2005, and in Bulgaria, Indonesia, Romania, Mexico and Ukraine in earlier years.

Source: James (2005 and earlier years)

Table 2.2 *Commercial production of GM crops worldwide, 1996 and 2005: Main crops*

	Total areas (million ha)			GM varieties as % of total for crop	
	1996	2005	% total 2005	1996	2005
Soy	0.5	54.4	60	<1	60
Maize	0.3	21.2	24	<1	14
Cotton	0.8	9.8	11	2	28
Canola	0.1	4.6	5	<1	18

Source: James (2005 and earlier years)

Table 2.3 *Commercial production of GM crops worldwide, 1996–2005: Main traits*

Traits	Crops	Area under cultivation (million ha)	Proportion of total area under cultivation (%)
Herbicide tolerance	Soy, maize, canola, cotton	63	70
Insect resistance	Cotton, maize	18	20
Stacked	Cotton, maize	9	10

Source: James (2005 and earlier years)

The most important crop is soy, followed by maize, cotton and canola. All these crops have been developed to carry the *herbicide tolerance* and/or the *insect resistance* trait. GM crops with an herbicide tolerant gene allow farmers to spray their fields killing weeds but not the crop under cultivation; the most common herbicide tolerant GM crop is soy, commonly called 'RR soy'. The main advantage is reduced chemical application. RR soy is often combined with 'no-till' cultivation, which reduces soil erosion. Anti-GM groups have raised concerns about the environmental risks of developing herbicide resistance in crops, saying that it might lead to 'superweeds', and about the socio-economic risk of decreasing employment opportunities for landless labourers, since this would be a labour-saving technology.

As explained in the US case study (see Chapter 3), private investments targeted economically important crops with large seed markets in the US, namely soy, canola, cotton and maize. These crops were developed to address productivity constraints faced by US farmers in the context of scarce labour and pressures to reduce chemical inputs. These new technologies lower costs (rather than increase yields) and have been particularly useful in improving farm incomes where farmers' costs for pesticides and weeding labour were high and savings could offset the higher price to be paid for seeds.

An insect resistant gene allows plants to produce their own insecticides; the most common insect resistant crops are *Bt* cotton and *Bt* maize, carrying genetic material from *Bacillus thuringensis (Bt)*. *Bt* cotton resists infestations of bollworms and *Bt* maize is effective against the European maize borer, both of which are common. The main advantage is reduced pesticide application, a health benefit saving farmers from pesticide poisoning. This is particularly important in developing countries where pesticides are applied by hand, often without protective gear. *Bt* cotton, in some cases, has shown higher yields. At the same time, in some developing country contexts farmers use fewer, if any, pesticides. Concerns have been raised about unintended impacts such as the build up of resistance in target insects – a great worry for organic farmers who use other management tools for insect control.

Hundreds of other crops and traits have been developed but are at different stages of the research-development-commercialization process. Dozens of crop varieties are under field trials and many have been approved for commercial release, from carnations, tomato, papaya and tobacco to sweet potato. Some are under commercial production but in very small quantities. Other desirable traits have also been developed. *Virus resistance* has been genetically engineered into tobacco, sweet potato and tomato, which allows farmers to reduce pesticide applications. Another aim of GM crop developers has been to develop *quality traits* for the consumer so that the products are healthier; golden rice that contains vitamin A is an important example. Other crops being developed include high-oleic soybeans, high-oleic canola/rapeseed and laurite canola, intended to deliver healthier oils in response to consumer demand for low cholesterol diets, especially in the US.

Thus, since the first generation GM crops, based on two events and four crops released in 1994, there has been no significant release of a major crop. The anticipated second generation GM crops are rice, followed by wheat, both major food crops. GM rice varieties have been developed and are at the final stages of regulatory approval in China (see Chapter 8) and were officially released in Iran in 2004 (James, 2005).

Technological innovation and institutional change

What explains this pattern of evolution? Why did the first products develop in the US and spread to those 13 countries? In their seminal works on agricultural growth, Hayami and Ruttan (1985; Ruttan, 2001) developed a theory of agricultural growth that focuses attention on the role of institutional change. They explain technological innovation as a process of interaction among four factors: factor endowments, institutions, technology and cultural endowments. Each of these four factors influences each of the others and is influenced by them; shifts in any of these factors can trigger a dynamic change. The technological development that takes place is thus dependent on institutional responses. In the case of GM crops, advances in the science of biotechnology – an exogenous factor – opened up new opportunities. The development of this technology into a commercial product and diffusion to farmers depends on institutional responses, in research and development and in seed marketing systems.

Institutional change for development and diffusion of GM crop varieties

Developing GM crops requires major institutional changes because the science of agricultural biotechnology is a significant departure from conventional breeding, requiring major investments in new capacity (see Box 2.1). In addition, the process of developing a GM crop, from laboratory experimentation to the farmer's field, is complex, involving not only a scientific process but also biosafety testing and commercialization.

As a scientific process, it involves:

- upstream scientific research – a wide range of research to build up knowledge on plant genetics;
- a transformation 'event' – a successful transfer of a gene from one organism to another. Once transferred, transgenes can be managed in the same way as genetic material in conventional cross breeding programmes;
- variety development – a successful development of a variety adapted to a local environment, using an adapted local variety.

As a biosafety testing process, it involves:

- a laboratory phase – the transfer of a gene takes place under carefully controlled laboratory conditions because of safety concerns. The GM plant is then grown in special glasshouses;
- a field testing phase – the plants are then grown in the open field. This requires authorization from a relevant national biosafety regulatory body for environmental release;
- approval for commercial release phase – commercial release is subject to authorization by a relevant regulatory body, based on an application documenting results of field trials and other data on environmental, health and in some countries, socio-economic risks.

As a commercialization process, it involves:

- seed multiplication – once a GM variety is approved for use by farmers for commercial production, it must be multiplied and either sold or given to farmers;
- licensing for intellectual property – products of research, the gene and variety are patented so that commercial multiplication can only be done under a licence negotiated with the patent owner;

- marketing to farmers – the variety is adopted by farmers who perceive benefits as the variety has characteristics such as higher yield, being more robust against disease or pests, performing better under drought, requiring less labour, fetching higher prices in the market and so on.

These processes are much more complex than for varieties developed by conventional breeding. They require different types of institutional arrangements, including: scientific and technical resources; biosafety management resources; intellectual property management systems. They require two important structural changes in the organization of R&D and seed markets:

1 *The creation of new scientific capacity in R&D in biotechnology* in addition to plant breeding, and arrangements for accessing proprietary technology held by the global private sector. Most developing countries have scientific capacity in plant breeding but few have capacity in biotechnology and molecular biology (Byerlee and Fischer, 2001).
2 *The creation of a regulated seed market in which farmers buy seed and only licensed sellers can sell seeds of authorized varieties.* This involves new legislation on biosafety and decisions on inclusion of plants for coverage under intellectual property legislation. More importantly, it requires a deeper institutional shift in the seed sector, which historically operates with multiple sources of supply: small commercial seed companies, state owned seed suppliers related to national research efforts, farmer to farmer sharing of well performing varieties, and farmers saving their own seeds. Obtaining good quality seeds of good varieties is something that farmers all over the world care about as a first step in managing their productivity.

These new institutional challenges are particularly challenging for developing country agriculture because considerable scientific, financial and administrative resources are required to make shifts. But institutional change is difficult in any society because it is a political process driven by interest groups that put pressure on government policy making, by public attitudes and by the behaviour of all stakeholders. Institutional shifts require mobilization of political resources by innovators (Hayami and Ruttan, 1985).

The conflicting interests of farmers who could benefit from GM varieties, scientists and industry as against non-GM farmers and environmental and anti-globalization groups, have made this technology one of today's most socially divisive issues, as each of the case studies in this volume demonstrates. The issue has pitted the EU against the US, unleashing trade disputes. But it has also divided societies. Some political economy analyses of the divisions focus on economic interests as driving stakeholder positions within countries and national positions internationally (Graff and Zilberman, 2004). Others focus on broader processes of politics at multiple levels – local, national and global – involving interactions among citizens, business, experts and the state in new ways (Herring, 2005; Scoones, 2006).

The key stakeholders in different countries include: farmers who would benefit from productivity increase (i.e. cotton, canola, maize and soy producers); seed companies producing or in line to produce GM varieties; the scientific community, which wants to be at the 'leading edge' of science and technology. It also includes those who would lose from technological change: farmers who have nothing to gain but are vulnerable to changes in ecosystems, such as indigenous people who are not growing any of the GM crops; and chemical firms and seed companies whose products would lose markets. Economic interests are not the only drivers of political pressure. Scientists are motivated to be at the forefront of their profession. Ideology and values are behind social movements for sustainable agriculture and environmental and consumer resistance to GM crops and anti-corporate globalization environmentalists have driven the opposition to them. The case studies in this volume show the path of development that this technology has taken in each of the countries, and the similarities and differences will be analysed here and in Chapters 11 and 12.

Why the US leads

US companies have responded to the new opportunities created by the invention of GM technology. In Chapter 3, Traxler argues that the large size of the domestic seed market was a critical factor in the market incentives for private sector investment. At the same time, the US government saw maintaining export competitiveness as a public policy priority. Both private and public sectors responded with new institutional arrangements.

First, private industry created new capacity to mobilize global science, invest in R&D for new GM varieties and market seeds to farmers. Chemical companies like Monsanto created this new capacity through licensing technology from universities and mergers and acquisitions of biotech and seed companies. The US government adopted pro-biotechnology policies, including sizeable funding for research. Universities began to license their research results to companies, facilitated by new legislation that permitted patenting of research financed by federal funds.

Second, biosafety regulations became progressively supportive, after the initially restrictive positions of the 1970s. The US government lifted restrictions on DNA research in the 1980s and in 1992 simplified the approval process (Osgood, 2001). The US regulatory process is 'permissive', with crops being screened for risks with the same procedures as for non-GM crops, in contrast with the 'precautionary' approach adopted in many other countries, which view GM crops as novel and different and screen for scientific uncertainties (Paarlberg, 2001).[3]

Third, changes in the intellectual property legislation were key to creating incentives for investing in R&D. Unless new varieties are patented the investor is not able to recover development costs because seeds propagate and can be easily reproduced, either by the farmer or by an entrepreneurial seed company.

This is why plant breeding has historically been dominated by the public sector, and why private sector activity was limited, mostly to hybrid varieties.[4] In the late 1980s and 1990s, however, a sea change occurred not only in the science of biotechnology but in intellectual property protection in the US that radically altered the incentive structure, progressively introducing stronger patents overall and stronger patenting of plants (Louwaars et al, 2005; Pardey and Wright, 2006). Up until the 1970s, plant breeding took place in an environment of weak patents in the US. Patents were limited to horticultural crops and to plant variety protection under Plant Variety Protection Certificates or 'Plant breeders rights' under the Union for the Protection of New Varieties of Plants (UPOV) convention. The latter gave 20 years of protection but allowed use of materials for breeding and in many cases permitted farmers to save seeds. The environment altered radically with the 1980 Diamond vs Chakravarty decision of the US Supreme court, which confirmed that plant varieties and genes could be patented. In the same year, the Bayh-Dole Act authorized patenting of federally funded research. This meant that non-profit and university researchers could patent their discoveries and license them to commercial investors, which made investments in biotechnology financially attractive.[5]

These institutional changes were critical to the development of the technology. The biotechnology industry then further lobbied for tighter intellectual property protection globally through the WTO and other trade agreements. Together with other knowledge-based industries (pharmaceuticals, information and communications technologies), the biotechnology industries were behind the Agreement on Trade-Related Aspects of Intellectual Property (TRIPS), which introduced minimum standards for all member countries of WTO. However, because plants and animals and 'essentially biological processes' can be excluded from patentability, not all countries have adopted patents for GM crops, and have retained flexibilities such as breeders' rights. This has affected the conditions of diffusion of GM varieties outside of the US.

Why Europe diverged

The evolution of GM technology in Europe, as described by Tiberghen in Chapter 4, contrasts sharply with its evolution in the US. Since the 1980s, this technology has been deeply divisive and controversial, leading to shifting policies in European countries. Currently, commercial production is negligible and imports of GM products are limited. A major social issue currently debated is to create 'GM free' consumer markets by requiring labelling, and 'coexistence' of GM, conventional and organic farming practices.

However, the initial reaction of European public policy, scientists, industry and farmers was positive. The European biotechnology industry responded to the scientific breakthroughs and began to invest heavily in parallel with the US. Public policy was also supportive into the mid-1990s, including public funding of research and granting approval of products for both local production and import. The EU approved five varieties of maize, one variety of soy and seven varieties of canola up to 2000. Over 200 field trials were being conducted annually up to 1999, focusing particularly on maize, canola and beets. The same economic interests that drove US investments were present. Like the US, Europe has large seed markets (see Table 3.5, Chapter 3), and policy makers saw biotechnology as important in maintaining the competitiveness of European agriculture.

But in the late 1990s, EU policies shifted and a series of regulations were legislated to severely limit the production and importation of GM products. These included a biosafety framework based on precautionary principles, a moratorium on new approvals and mandatory labelling of products with GM crop content.

Some argue that these policies responded to protectionist interests (Graff and Zilberman, 2004; Anderson and Jackson, 2006). Tiberghien argues that they went against economic interests of major stakeholders (see Chapter 4). The key factor was not economic interest but public rejection and a dramatic shift in public opinion around 1999, driven by the rise of opposition led by environmental and health NGOs in 1996, and later joined by anti-globalization groups, allied with green parties where they existed, as well as the European Parliament. These restrictive policies undermined Europe's biotechnology industry, which includes some of the leading multinationals such as Syngenta, Aventis and Bayer, and numerous smaller firms. The countries with the strongest opposition are also home to these firms (Switzerland for Syngenta) and numerous other ones (France and Germany). Restrictive legislation may also have undermined the interests of farmers and the livestock industry; France, Germany and the UK are large producers of crops like canola and maize and the livestock industry depends on imports of feed.

Why was the civil society opposition so strong in Europe and weak in the US? I would suggest that what matters are not just the economic interests of both consumers and producers but also cultural values and social movements. Tiberghien points out that Greenpeace and others leading the anti-GM campaign were surprised by the public support for their movement. GM technology is symbolic of Europe's weakness in managing globalization, a process dominated by the US. The strength of public hostility was also thought to be related to the general mistrust of science and scientists, generated by the public mishandling of mad cow disease when government scientists initially denied the risks. GM technology is also symbolic of the type of agriculture that many in Europe want to turn away from: science- and industry-led activity aimed at production. Indeed, key objectives of agricultural policy have little to do with productivity but emphasize environmental sustainability and lifestyle preservation. For example, Swiss agricultural policy is concerned with preserving tall trees in the landscape and maintaining farming communities that would otherwise disappear.

Why West and Central Africa have not yet adopted the new technology

In contrast to both the US and Europe, there has been no commercial production of GM crops nor significant R&D in West and Central Africa. This is not because there was no demand from scientists and farmers, but because of overwhelming constraints in both market incentives and capacity.

The case study by Nwalozie and others (see Chapter 5) explains that policy makers in the subregion see the potential of the technology to respond to pressing public priorities of improving food security and farm incomes. Yet the size of the seed market is too small to be attractive to private investors and the capacity of the public sector is too weak to respond.

Though the public priority is to address major food crops like cassava and millet, can West and Central Africa make use of the technology that has been developed globally and benefit from the GM varieties already developed? Maize, canola and soy are not grown in this region but cotton is a major crop. For some countries, it is the primary source of foreign exchange. Mali, Côte d'Ivoire, Benin and Burkina Faso ranked 13th, 15th, 16th and 17th among world producers of cotton in 2004. Monsanto has started a programme of trials in Burkina Faso but this is still in the preliminary stage. Diffusion of this technology still confronts the constraints of institutional capacity in adaptive R&D, as well as in the creation of a regulated seed market. Research institutions are weak and seed markets are informal.

Why the crops spread globally to Canada, Argentina, Brazil, China, India and South Africa but not elsewhere

The first generation of GM varieties were developed initially for US farmers. Canada, Argentina and China were the first to adopt these varieties, and in the case of China, develop their own versions. Diffusion has been rapid in Brazil and India in the last few years, though reliable data for earlier years are not available because the varieties were not officially released (see Chapters 6, 9, 11). Diffusion in other countries, which started in the 1990s, has been slow to increase or has petered out.

As in the US and Europe, these countries had significant economic interest in adopting the GM varieties that had been developed as they are the main producers and exporters of soy, cotton and maize. Canada is a major producer of canola. Policy makers saw these crops as having strategic national interest for expanding and maintaining export competitiveness, with farmers benefiting financially. For the multinational seed companies, these countries presented an important market opportunity for their products.

Table 2.4 shows the countries that together account for 90 per cent of total global production for each of the four crops. Not surprisingly, the GM producing countries are among the top global producers of four GM crops:

Table 2.4 *Top producing countries of GM and conventional soy, maize, cotton and canola, 2005 (country and million tonnes produced)*

	Soy	Maize	Cotton*	Canola
Global production ranking	209.5	692	202	45
1	US (82.8)	US (280)	China (50.9)	China (11.3)
2	Brazil (50.2)	China (131)	US (36.4)	Canada (8.4)
3	Argentina (38.3)	Brazil (34.9)	India (26.4)	India (6.2)
4	China (16.9)	Mexico (20.5)	Pakistan (18.9)	Germany (4.7)
5	India (6)	Argentina (19.5)	Uzbekistan (10.3)	France (4.4)
6		India(14.5)	Turkey (9.5)	UK (1.9)
7		France (13.2)	Brazil (8.0)	Poland (1.4)
8		Indonesia (12.0)	Australia (5.9)	Australia (1.1)
9		South Africa (12)	Greece (3.4)	Austria (0.9)
10		Italy (10.6)	Egypt (2.8)	Czech Republic (0.8)
11		Romania (10)	Syria (2.50)	U.S. (0.7)
12		Hungary (9)	Mali (2.2)	
13		Canada (8.3)	Turkmenistan (2.0)	
14		Ukraine (7.2)	Côte d'Ivoire (1.7)	
15		Egypt (6.8)	Benin (1.5)	
16		Serbia and Montenegro (6.3)	Burkina Faso (1.5)	
17		Philippines (5.2)	Tajikistan (1.2)	
18		Nigeria (4.8)		
19		Thailand (4.2)		
20		Spain (4.0)		
21		Germany (3.8)		
22		Russian Federation (3.7)		
23		Viet Nam (3.5)		
% of total world production	93% in 5 countries	90% in 23 countries	91% in 17 countries	92% in 11 countries

Note: *data for 2003.

Source: FAO (2004); Cotton data from Bulletin of International Cotton Advisory Committee (www.icac.org).

- Soy: the US, Brazil and Argentina are the top three producers in the world, together accounting for 82 per cent of global production in 2005. These were also the first countries to adopt RR soy.
- Cotton: China, the US and India are the top three producers, accounting for 56 per cent of global production in 2003. These were also the first countries to adopt *Bt* cotton.
- Maize: global production is less concentrated, and GM varieties have not spread to as many countries. The top 10 producing countries are the US, China, Brazil, Mexico, Argentina, India, France, Indonesia, South Africa and Italy. Of these, only in the US are GM varieties grown extensively. In Argentina and South Africa they were introduced but did not diffuse rapidly. GM maize is concentrated in the US and Canada, which are respectively placed 1st and 13th in production.
- Canola: China, Canada, India, Germany, France, UK, Poland, Austria and the Czech Republic are the top ten producers but GM varieties are only grown in significant areas in Canada and in the US, which ranks 11th in global production.

Why did these varieties not extend to some of the major producing countries? A part of the reason is agronomic: the varieties were not appropriate for the local growing conditions, or for the local factor price environment. But institutional factors have also been important. GM varieties did not spread to the major canola producing countries because they are European countries that took anti-GM positions as a matter of policy. Alternatively, it is less obvious why China and India did not adopt GM canola, GM soy or GM maize. This will be explored in Part III of this volume. It is also surprising that GM varieties of cotton have not been diffused substantially in Pakistan, Uzbekistan, Turkey and Australia. These questions need to be investigated further for a fuller understanding of the institutional factors shaping the spread of GM crops.

Public policy adjustments were also a factor in the global marketing of GM seeds. A global shift to tighter intellectual property rights took place during the 1990s, through TRIPS. TRIPS is one of the WTO agreements that are binding on all member countries. US-based multinationals in the large knowledge-based industries – pharmaceuticals, agricultural biotechnology and computer industries – were heavily engaged in lobbying for TRIPS and continue to promote strong intellectual property provisions in trade agreements. These provisions make global seed markets commercially attractive to multinationals with proprietary technology.

Alternative institutional models

The institutional environment that produced and spread the first generation of products was the corporate model, with the exception of developments in China. This environment offered incentives of large markets and tight

Table 2.5 *Institutional approaches to development of GM crop varieties*

Investing in GM varieties	US model	Possible alternatives
Investing in GM varieties		
R&D – upstream biotechnology research: finance	Public sector Private corporate sector Venture capital	Public Private corporate Venture capital *Global public or non profit (development aid)?**
R&D – GM crop variety development: biotechnology step	Public sector Private corporate sector Biotech start ups	Public NARs? Private multinational? Biotech startups *Global public or non profit (CGIAR)?* *Public private partnerships?*
R&D – GM crop variety development: plant breeding step	Large multinationals	Joint venture multinational/local National private seed companies *Public NARs?* *Public/private partnerships?*
Commercialization – approval process, intellectual property rights	Large multinationals	Joint venture PPPs National private seed companies? National NARs
Financing R&D for commercial product development	Patent protection and collection of technology fee	Govt. budget allocation for national priority *International budget allocation for global public good?*
Intellectual property rights	Strong intellectual property right protection on genetic resources – licence required for use in research and for product development. Costs of licensing reflected in higher costs of seeds.	Breeders' rights No intellectual property rights on genes and plant varieties
Regulated seed market for GM varieties		
Seed – Biosafety approval	Permissive	Range of choice from permissive to precautionary
Enforcement of biosafety approval	No active monitoring by government. But all farmers buy in commercial seed market	Government monitoring of farms Civil society watchdogs Incentives for informal seed supplies to go through process

Seed supplier – licensed	Large seed companies (seed companies merged with agro-chemical companies)	Multinationals in partnership with local seed company Informal sector suppliers: farmer saved, farmer to farmer, small private entrepreneurs, small state enterprises
Enforcement	No active monitoring by government Farmers contract not to save seeds Legal suits by intellectual property owner in case of infringement	Monitoring by government Farmers contract not to save seeds? Civil society watchdogs
Collecting the technology fee	In price of seed	In price of seed At sale of crop? At export of crop? No collection?

Note: *Italicized approaches are not currently implemented and may not be feasible or justifiable.

intellectual property protection on genes and seeds that could mobilize the large magnitude of finance needed to invest in R&D. It also offered a seed market that was commercialized with a supportive regulatory system in which farmers, seed companies and consumers had basic confidence. Institutional innovations that took place involved adjustments in the corporate structures where agro-chemical companies were transformed into agro-chemical/seed conglomerates, and where permissive biosafety and stronger intellectual property legislation were pushed forward and globalized to reach global markets.

Is this the only institutional environment in which agricultural biotechnology can be applied to develop useful varieties for farmers and where risks can be managed? There are alternative ways of providing incentives, sourcing finance and making seed markets work. As illustrated inTable 2.5, the public sector in developing countries would have a much greater incentive where investing in improving agricultural productivity is a more important national priority because of the size of the population and agriculture's importance in the GDP. While the size of the seed market may drive private sector investments, public sector investments should be driven by the size of the population who would benefit and the size of the aggregate economic gain. Where the seed market would need to be regulated, farmers purchasing from large seed companies and entering into a contractual agreement not to save seeds for replanting need not necessarily be the only means for supplying seeds, collecting the technology fee or enforcing biosafety.

This model has favoured the development of crops with large markets in the US as well as access to these crops in developing countries for which they are suited. It is not necessarily the model that would be adequate for developing priority crops in poor developing countries as the West/Central African experience illustrates. Nor is it necessarily the model that would be most appropriate for the use of this technology in developing countries in general. The rest of this volume reviews the experiences of countries where GM crop varieties are being commercialized, showing how the new technology is being acquired and developed on the one hand, and how seed markets are adapting on the other.

Notes

1 Jikun Huang, correspondence, May 2006.
2 June, 2006.
3 See Paarlberg (2001) for a comprehensive review and typology of government policy approaches ranging from 'promotion', 'permissive', 'precautionary' and 'restrictive'.
4 See Beintema and Pardey (2001) for a comprehensive review of agricultural research over the last decades including issues of the shifting roles of private and public sectors.
5 See Parday and Wright (2006) for a detailed explanation of the evolution of intellectual property for plant agricultural biotechnology and its impact on R&D investments.

References

Anderson, K. and Jackson, L. A. (2006) 'Transgenic crops, EU precaution, and developing countries', *International Journal of Technology and Globalisation*, vol 2 (1/2), pp65–80

Byerlee, D. and Fischer, K. (2001) 'Accessing modern science: Policy and institutional options for agricultural biotechnology in developing countries', *World Development*, vol 30 (6), pp931–948

Cohen, J. (2005) 'Poorer nations turn to publicly developed GM crops', *Nature Biotechnology*, vol 23 (1), www.nature.com/nbt/index.html

FAO (2004) *State of Food and Agriculture 2003–04 – Agricultural Biotechnology: Meeting the Needs of the Poor?* FAO, Rome

Graff, G. and Zilberman, D. (2004) 'Explaining Europe's resistance to agricultural biotechnology', *Update, Agricultural and Resource Economics*, vol 7 (5), University of California, Giannini Foundation of Agricultural Economics, Berkeley

Hayami, Y. and Ruttan, V. (1985) *Agricultural Development, An International Perspective*, revised and expanded edition, Johns Hopkins University Press, Baltimore

Herring, R. (2005) 'Miracle seeds, suicide seeds and the poor: GMOs, NGOs, farmers and the state' in Raka, R. and Katzenstein, M. (eds) *Social Movements in India: Poverty, Power and Politics*, Rowman and Littlefield, Lanham

James, C. (2005) *Global Status of Commercialized Transgenic Crops: 2005*, International Service for the Acquisition of Agri-Biotech Applications (ISAAA), Manila

Louwaars, N. P., Tripp, R., Eaton, D., Henson-Apollonio, V., Hu, R., Mendoza, M., Muhhuku, F., Pal, S., and Wekundah, J. (2005) 'Impacts of strengthened intellectual property rights regimes on the plant breeding industry in developing countries: A synthesis of five case studies', Report commissioned by the World Bank, Waningen UR, Centre for Genetic Resources, Waningen, The Netherlands

Mokyr, J. (2002) *The Gifts of Athena: Historical Origins of the Knowledge Economy*, Princeton University Press, Princeton

Nelson, R. (1994) 'Economic growth via the coevolution of technology and institutions', in Leydensdorff, L. and Van den Besselaar, P. (eds) *Evolutionary Economics and the Chaos Theory: New Directions in Technology Studies*, Pinter Publishers, London, pp21–32

Osgood, D. (2001) 'Dig it up: Global civil society's response to the plant biotechnology agenda', in Anheir, H., Glasius, M, and Kaldor, M. (eds) *Global Civil Society, 2000*, Oxford University Press, Oxford, pp79–107

Paarlberg, R. (2001) *Politics of Precaution*, Johns Hopkins University Press, Baltimore

Pardey, P. and Wright, B. (2006) 'The evolving rights to intellectual property protection in the agricultural biosciences', *International Journal of Technology and Globalisation*, vol 2 (1/2), pp12–29

Pray, C. and Naseem, A. (2003) 'The economics of agricultural biotechnology research', ESA Working paper No. 03-07, FAO, Rome, www.fao.org/es/esa

Ruttan, V. (2001) *Technology, Growth and Development: An Induced Innovation Perspective*, Oxford University Press, Oxford

Scoones, I. (2006) *Science, Agriculture and the Politics of Policy: The Case of Biotechnology in India*, Orient Longman, Hydrabad

3

US: Leading Science, Technology and Commercialization

Greg Traxler[1]

Introduction

The emergence of practical biotechnology protocols for creating GM crop varieties has transformed the system for supplying improved varieties to US farmers. Compared to nearly any previous agricultural technology, adoption of GM varieties in the US has been rapid. Five GM varieties were introduced to the market in 1996. By 2005, GM varieties occupied 87 per cent of the total soybean area, 79 per cent of the cotton area and 52 per cent of the maize area in the US. Within three years of the appearance of the first GM varieties in the US more than 500 such varieties were available for field crops.

US government policy has strongly favoured the use of GM varieties. Significant levels of public funding for biotechnology research and for maintaining the required administrative infrastructure have been committed. The federal government, through the Agricultural Research Service (ARS) of the US Department of Agriculture (USDA), spends well over US$120 million per year on plant biotechnology, and public and private universities provide substantial additional funding. The US is a major exporter of the main GM crops, and the US advantage in biotechnology is seen as a simple extension of its strong existing agricultural science base. A quarter or more of the soybean, corn and cotton crop is exported to world markets and reducing production costs through better technology is a tried and true means of maintaining the country's competitive advantage. The government's strong public support for biotechnology has been relatively uncontroversial. A few civil society groups have expressed concern about the high level of GM variety production but their voices have had little or no effect on the adoption and use of GM crops.

As GM commodity production is not segregated from conventional pro-duction at any point along the marketing chain, virtually all US exports contain GM material. Because of the importance of exports to the agricultural sector, the US has strongly resisted international attempts to restrict trade in GM commodities, recently opposing international labelling requirements. Three institutional structures are crucial for a country to access GM technology. First, the scientific capacity in biotechnology and crop improvement must exist. Sec-ond, a regulatory system that has the confidence of industry and consumers must be in place. Finally, there must be a dynamic system of seed delivery. This chapter reviews the US experience with GM varieties and discusses the fac-tors that have contributed to the success of the US biotechnology movement. Activity within the US scientific, regulatory and commercial seed sectors is examined and data on the origin of GM events and their movement into com-mercial crop varieties traced.

Crop improvement research in the US

The US seed market is supported by a large crop improvement research ef-fort shared by three sets of institutions: the ARS/USDA, state agricultural experiment stations (SAESs) and private companies. Approximately 2,023 crop improvement scientists were employed in the US in 2001 (see Table 3.1). Bio-technology research has become an important research focus, with over 33 per cent of all US crop improvement researchers conducting biotechnology related research. The private sector employs 70 per cent of all biotechnology scientists, the SAESs, 18 per cent, and ARS/USDA, about 12 per cent (see Table 3.2). This demonstrates the vital role that the private sector has played in the de-velopment of US seed and biotechnology research. The difference in intensity of private sector research between the US and other countries is one of the primary reasons that the US has led the way in the use of GM technology.

Regulation of GM varieties in the US

The regulation of GM varieties in the US is a coordinated effort shared by four principal agencies: the USDA's Animal and Plant Health Inspection Service (APHIS) regulates the field testing of GM varieties; the Food and Drug Adminis-tration (FDA) governs food safety and labelling; the National Institutes of Health employ guidelines developed for the laboratory use of genetically engineered or-ganisms; and the Environmental Protection Agency (EPA) ensures the safety and safe use of GM varieties in the environment.

APHIS has overall responsibility for protecting agriculture from all types of pests and diseases. This includes regulating 'the import, handling, interstate movement, and release into the environment of regulated organisms that are

Table 3.1 *Numbers and percentages of plant breeding science years devoted to plant breeding research, germplasm enhancement, cultivar development and biotechnology in SAESs, USDA (including Agricultural Research Service and plant materials centres) and private industry, 2001*

Category	SAESs		ARS/USDA		Private industry	
Plant breeding research	85	20% ⎤			180	13%
Germplasm enhancement	70	17% ⎬	138	64%	96	7%
Cultivar development	144	34% ⎦			632	46%
Biotechnology R&D	121	29%	80	36%	476	34%
Totals	420	100%	218	100%	1385	100%

Source: Author's survey

Table 3.2 *Private and other sector breeders as a percentage of total plant breeders by research area*

	SAESs	ARS	Private	Total
Conventional plant breeding	22%	10%	68%	100%
Biotechnology R&D	18%	12%	70%	100%

Source: Author's survey

products of biotechnology, including organisms undergoing confined experimental use or field trials' (USDA, 2005). APHIS oversees GM crop field trials and publishes information on articles that have been field tested, some of which are summarized below. The EPA also regulates the sale, distribution and use of pesticides, including biological pesticides produced using biotechnology. Developers must obtain an Experimental Use Permit from the EPA before field testing a GM variety and the EPA can place restrictions on commercialization and use of any GM product. The FDA also has responsibility for the safety and proper labelling of all plant-derived foods and feeds in the US. The FDA testing and approval process does not distinguish between foods that are imported or domestically produced, or between products produced using conventional plant breeding or through biotechnology techniques. Any food additive,[1] including those introduced though plant breeding, must receive FDA approval before commercialization.

Two main steps are involved in obtaining APHIS clearance for a GM variety for commercial use. The institution producing a new GM variety must first either obtain a permit to conduct field trials or notify APHIS of its intent to conduct field trials of the new 'regulated article'.[2] Upon completion of several years of field trials, the institution may petition APHIS to have an article removed from regulated status. If the petition is granted, the GM variety may be commercialized. Once an article is removed from regulated status, subsequent GM varieties of the same crop can be developed without additional approvals if they are produced using conventional plant breeding techniques.

A total of 10,880 trials were conducted by 242 different institutions in the US between 1987 and May 2005 (see Figure 3.1). An average of 997 trials was conducted annually between 1998 and 2004. The 40 institutions with the most field trials, reported in Table 3.3, include 12 public and 28 private institutions. Monsanto has conducted 40 per cent of all trials and received nearly a third of the 65 commercial approvals. An average of five commercial approvals per year has been granted since 1992 (see Table 3. 4). It is not clear how many different events have been field tested but a guess would be that less than one in 200 tested events has been approved for commercialization. Twenty-one different private companies have received at least one APHIS commercialization approval but the large life science firms dominate; only Monsanto, AgrEvo, Dow and Calgene have received more than two approvals and these four companies account for two-thirds of all approvals. Just two approvals have been received by public sector institutions (virus resistant papaya (Cornell University), and sulfonylurea tolerant Flax (University of Saskatchewan). Only papaya has been approved for commercialization and is now successful in Hawaii)). So while the capacity to generate a GM variety and conduct field trials is widespread, the commitment and financial muscle needed to obtain commercial approval is concentrated in large corporations. These firms possess expertise in

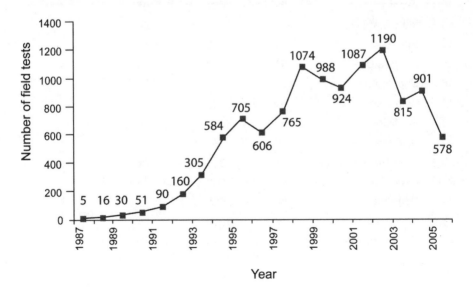

Figure 3.1 *Number of field tests in US by year (1987–May 2005)*
Source: USDA/APHIS webpage accessed at http://www.isb.vt.edu/2002menu/regulatory_ information.cfm, May 20, 2005

negotiating the regulatory process and dealing with post-approval intellectual property rights (IPRs) and have an ability to handle marketing challenges that public sector and smaller private sector companies lack.

Table 3.3 *Numbers of field trials and commercial approvals of the 40 institutions with the most field trials, 1987–May 2005*

Rank	Institution	Number of trials	Cumulative number of trials	Cumulative % of trials	Number of approvals
1	Monsanto	4396	4396	40	21
2	Pioneer	633	5029	46	1
3	AgrEvo	326	5355	49	9
4	Du Pont	320	5675	52	2
5	ARS	249	5924	54	
6	Seminis	199	6123	56	
7	Syngenta	182	6305	58	
8	DeKalb	181	6486	60	2
9	Calgene	164	6650	61	9
10	Scotts	150	6800	63	
11	Dow	132	6932	64	4
12	Aventis	122	7054	65	1
13	Iowa State U.	115	7169	66	
14	ArborGen	104	7273	67	
15	Rutgers U.	97	7370	68	
16	U. of Idaho	96	7466	69	
17	Betaseed	94	7560	69	
18	ProdiGene	90	7650	70	
19	DNA Plant Tech	89	7739	71	1
20	Stine Biotechnology	86	7825	72	
21	U. of Florida	83	7908	73	
22	Northrup King	80	7988	73	1
23	Novartis Seeds	77	8065	74	2
24	U. of Kentucky	75	8140	75	
25	Asgrow	75	8215	76	1
26	Upjohn	73	8288	76	1
27	U. of Nebraska	69	8357	77	
28	Cargill	65	8422	77	

29	Oregon State U.	65	8487	78	
30	Agracetus	61	8548	79	
31	Harris Moran	61	8609	79	
32	U. of Arizona	56	8665	80	
33	Stanford U.	54	8719	80	
34	Frito Lay	54	8773	81	
35	Agritope	53	8826	81	1
36	Michigan State U.	51	8877	82	
37	North Carolina State	50	8927	82	
38	Zeneca	50	8977	83	1
39	PetoSeed	50	9027	83	
40	Bayer CropScience	46	9073	83	

Source: Information Systems for Biotechnology (2005)

Table 3.4 *Number of commercial approvals in the US, 1992–2004*

Year	Number of Approvals
1992	1
1993	0
1994	6
1995	12
1996	10
1997	6
1998	9
1999	6
2000	2
2001	1
2002	5
2003	2
2004	5
Total	65

Source: Information Systems for Biotechnology website, www.isb.vt.edu/2002menu/regulatory_information.cfm

The US seed market

The US has been the focus of agricultural biotechnology research primarily because of the huge size of its domestic seed market. Its commercial seed market is larger than the size of the next two largest seed markets combined (see Table 3.5). Because of the value-added that they provide to seeds, the entrance of GM products has greatly increased the total value of seed sales in the US. For example, in 2005 Delta and Pine Land (D&PL) conventional varieties sold for US$26.12 per acre while the most popular GM varieties were priced from US$43.09 to US$61.41 per acre. This provides a powerful incentive for private sector investment in plant breeding research. Worldwide an estimated 70 per cent of biotechnology investments originate in the private sector (see Table 3.6) so this incentive is a key factor explaining the industry's growth and focus. The US market accounts for more than 85 per cent of worldwide revenues from licensing of GM technologies (see Figure 3.2).

Table 3.5 *Estimated size of domestic market for seed and other planting material, selected countries, 2005 (US$ million)*

Country	Size of domestic market	Country	Size of domestic market
USA	5700	Egypt	140
China	3000	Belgium	130
Japan	2500	Chile	120
France	1930	Serbia & Montenegro	120
Brazil	1500	Nigeria	120
Germany	1000	Finland	103
India	1000	New Zealand	90
Argentina	930	Slovakia	90
Italy	780	Switzerland	80
Canada	550	Paraguay	70
Russian Federation	500	Tunisia	70
Republic of Korea	400	Uruguay	70
Australia	400	Bangladesh	60
Mexico	350	Portugal	60
Taiwan	300	Ireland	60
Spain	300	Israel	50
Poland	260	Kenya	50
Czech Republic	200	Colombia	40
United Kingdom	257	Bolivia	35
Turkey	250	Zimbabwe	30
Netherlands	208	Peru	30
South Africa	217	Slovenia	30
Hungary	200	Saudi Arabia	18
Denmark	170	Zambia	15

Austria	170	Ecuador	12
Morocco	160	Malawi	10
Sweden	155	Dominican Republic	7
Greece	140	Uganda	6

Total = 25,243 *

Note: *This total represents the sum of the commercial seed markets of the listed countries. The commercial world seed market is assessed at approximately US$30 billion.
Source: International Seed Federation (2005)

Table 3.6 *Estimated global R&D expenditures on crop biotechnology, 2001*

	US$ millions	
Private (70%)	3100	
Public (30%)	1120	
Industrial country total (96%)		**4220**
China	115	
India	25	
Brazil	15	
Others	25	
Developing country total (4%)		**180**
World total		**4400**

Source: James (2002)

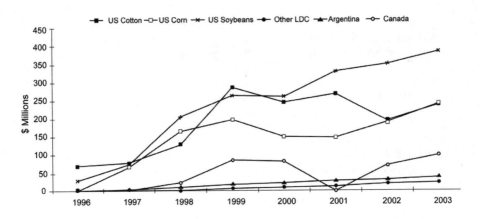

Figure 3.2 *Estimated royalty revenue from GM varieties in US and elsewhere*
Source: Author's calculations

To understand the effect that GM varieties have had on the cotton seed industry it is useful to examine data published by the USDA on the area covered by each cotton variety. The first commercial GM cotton varieties were developed through a strategic alliance between Monsanto and the dominant US seed cotton firm, D&PL. Monsanto chose to introduce *Bt* cotton through a D&PL licensing agreement. Initially D&PL was the only seed company to sell GM varieties. The elite commercial germplasm for the GM varieties was provided by D&PL from two recurrent parent varieties that were popular commercially in the US. The first US *Bt* varieties, NuCOTN 33[B] and NuCOTN 35[B], were subsequently marketed in several countries without in-country adaptation. In 1996, the first year of commercial availability, *Bt* cotton was planted on 729,000 hectares or 14 per cent of the cotton area in the US.

The introduction of GM varieties in 1996 led to a slight increase in the already-dominant cotton seed market position (based on area) of D&PL (see Figure 3.3). However, D&PL's market position has fallen to about 50 per cent as other seed companies have added GM varieties to their product lines. The number of GM varieties available increased steadily over time (see Figure 3.4) until, in 2004, it exceeded the number of conventional varieties.

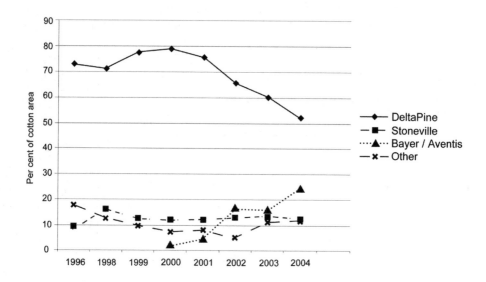

Figure 3.3 *US cotton seed market shares by seed company*
Source: USDA/AMS "Cotton Varieties Planted"

Table 3. 7 *Number of cotton varieties, total area and average area per variety by trait, US, 1996–2004*

Year	B2	Bt	Bt + RR	B2 + RR	BXN	RR	All GM	Tot. conv.	Total
				Number of varieties available					
1996		2	0		2	0	4	NA	NA
1997		12	3		2	12	29	NA	NA
1998		17	12		3	15	47	127	174
1999		19	17		3	18	57	116	173
2000		13	20		3	17	53	96	149
2001		10	24		4	28	66	99	165
2002		11	22		5	30	68	88	156
2003	1	9	22	2	4	32	70	77	147
2004	1	7	25	7	3	41	84	70	154
				Total area (1000ha)					
1996		697	0		8	0	705	5123	5828
1997		986	24		63	184	1257	4269	5526
1998		1042	192		315	926	2475	2906	5381
1999		1626	0		NA	2288	3914	2304	6218
2000		944	1259		NA	1637	3840	1900	5740
2001		212	1985		213	1667	4077	1230	5307
2002		637	1078		142	1764	3620	1279	4899
2003	0	754	1454	10	25	1724	3967	1420	5386
2004	0	918	1722	81	13	1722	4456	1283	5739
				Avg. area/variety (1000ha)					
1996		349	–		4		177	NA	NA
1997		82	5		32	15	43	NA	NA
1998		58	14		91	64	51	22	
1999		49	52		NA	67	61	21	
2000		49	82		NA	101	83	15	
2001		21	83		53	60	62	12	
2002		58	49		28	59	53	15	
2003		84	66	5	6	54	57	18	
2004		131	69	12	4	42	53	18	

Note: B2 = Bollgard II, *Bt* = *Bacillus thuringiensis kurstaki* insect resistance, RR = Roundup Ready, *Bt*+RR = Stacked, BXN = bromoxynil tolerant, NA = information not available.
Source: USDA/AMS (various years)

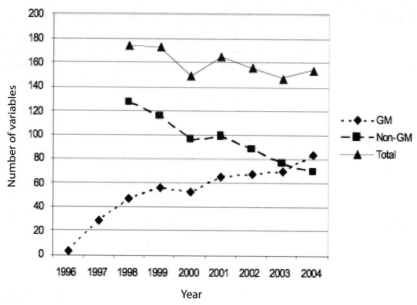

Figure 3.4 *Number of cotton varieties available in the US, 1996–2004*
Source: USDA/AMS "Cotton Varieties Planted"

By 2004, US farmers could choose from more than 84 different GM cotton varieties marketed by nine different companies (see Table 3.7). The average area per GM variety is not large (just 53,000 ha) and has been relatively steady over time. This suggests that the marginal cost of developing a new variety is not high and has important implications for the spread of GM technology to smaller countries.

A model of the transfer of GM technology to developing countries

How might smaller countries gain access to GM varieties? The US, with nearly 100 million hectares in major row crops, represents a market worthy of large investments, even under uncertain conditions. Under what conditions might other countries exploit spillover opportunities in biotechnology? A cost function model of the variety production process can be used to analyse a seed firm's decision to attempt to enter a developing country market, assuming that the firm is not currently doing business in that country. The firm will not enter a new market unless the expected market price for the new variety exceeds the expected average production cost. Figure 3.5 uses average cost curves to represent the market entry decision for three types of GM varieties:

1   a conventional (non-GM) variety being developed from available genetic resources;

2 a GM variety being developed by backcrossing a gene from an unadapted US variety to a locally available conventional variety, using a national licensing partner;

3 a directly marketed US GM variety, as was the case with the first cotton *Bt* variety, DP 33B.

Developing a competitive conventional variety would be the most expensive undertaking; importing a US GM variety would be the least expensive. Launching a conventional breeding programme implies substantial fixed costs associated with establishing a research facility, assembling a germplasm collection and hiring personnel. Depending on the quality of the initial breeding germplasm, it could take a minimum of five to ten years to get a product to market and, when vying for a share of the conventional seed market, success is quite uncertain. This explains why private investment in plant breeding research is so small in developing countries.

The limited experience to date with the launch of GM varieties in new markets has been one of relatively rapid capture of market share. GM variety development costs are quite low compared to those of conventional varieties because the GM variety is developed by backcrossing a single gene into a successful existing variety whose agronomic properties have already gained market acceptance. Fixed costs are very low for the biotechnology firm since it is assumed that it is partnered with a national seed company.

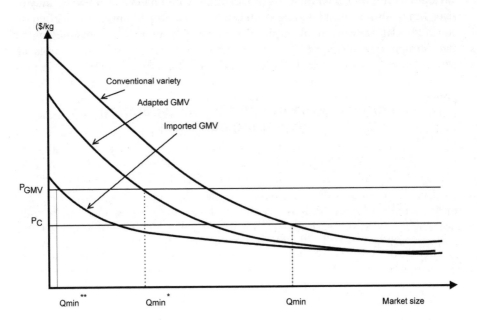

Figure 3.5 *Average cost curves and minimum market size required for market entry for conventional varieties, adapted GM varieties and imported GM varieties*

Where an existing US-developed GM variety is directly introduced into a new market, virtually no fixed costs are incurred since no breeding is done. Seed need only be tested, reproduced and marketed. Monsanto has used this type of commission agreement with seed distributors in South Africa, Argentina, Mexico, Australia and other countries.

Within this simple framework, it is clear that the minimum market size needed for a country to entice entry by a multinational biotechnology firm is smaller for GM varieties than for conventional varieties. Not only are R&D costs lower, but GM varieties generally sell at a premium of US$15 to US$35 per hectare. *Bt* cotton seed sells for double the price of conventional varieties in the US. Through the use of a licensing agreement, this premium will be shared between the biotechnology firm and the licensee seed company.

In Figure 3.5, the minimum size for entry would be Q_{min} for a conventional variety, Q_{min}^* for an adapted GM variety and Q_{min}^{**} for an imported GM variety. This implies, for example, that Pioneer or Asgrow might choose to enter into GM variety partnerships to license their genes to seed companies in countries where they have not tried to breed their own conventional varieties. Because of the price premium on GM variety seed, it is also possible that multinationals might choose to enter markets in countries even if it means investing in setting up a conventional breeding programme. Development costs are no higher for a GM variety than for a conventional variety once the genetic event has occurred.

The above discussion applies to giving developing countries access to GM varieties containing genes already in use in the US. The events involved may or may not be useful in addressing problems that are important in tropical countries. The R&D investment required to develop a useful article is far higher than the investment required for adapting an existing one to local germplasm. Multinational companies will invest in biotechnology research for some important developing country problems only if those countries represent an attractive potential market. In general, the attractiveness of developing countries for the introduction of GM varieties depends on the applicability of the technology as well as the marginal costs of entering new markets. To date, the successful GM technologies have been highly transferable, more similar to pesticides than to new plant varieties. The potential to amortize large research investments over markets in several countries provides an incentive for firms to undertake them.

Even if a country does not have the scientific, regulatory and commercial institutional capacity to develop its own GM varieties it can gain access to spillovers from GM varieties developed for other markets. This requires a much less daunting set of institutional capacities. The most important conditions might include:

- The country must have a tested, transparent, science based regulatory process for GM varieties;
- There must be a reasonable ability to protect intellectual property embodied in seed;

- GM varieties should be accepted by farmers, regulators, processors and legislators;
- Seed markets must be capable of delivering seed to farmers.

Summary

The US has moved more quickly to acceptance of the use of GM crop varieties than other countries. The rapid market penetration of GM varieties in the US is based on the existence of a large, dynamic seed market, an experienced and predictable regulatory system and a significant public and private sector research capacity. A few large developing countries, including Brazil and India, are developing their capacity in these areas. A smaller set of institutional capacities is required for a country to access spillover benefits from GM varieties developed in other countries.

Notes

1   Food additives are 'substances introduced into food that are not pesticides and are not generally recognized as safe by qualified scientific experts'. GM products are pre-screened by FDA to decide whether they contain a 'food additive'. Only one GM crop derived product has been judged to contain a food additive (USDA, 2005).
2   'Regulated articles are considered to be organisms and products altered or produced through genetic engineering that are plant pests or that there is reason to believe are plant pests' (APHIS, 2006).

References

APHIS (2006) 'United States regulatory oversight in biotechnology responsible agencies – Overview'
Information Systems for Biotechnology (2005) 'Biotec crop database', available at www.isb.vt.edu/2002menu/regulatory_information.cfm, accessed 21 May 2005
International Seed Federation (2005) 'Seed statistics: World seed trade statistics', available at www.worldseed.org/statistics.htm, accessed 21 May 2005
James, C. (2004) 'Global status of commercialized biotech/GM crops: 2004', ISAAA Briefs No. 32-2004, The International Service for the Acquisition of Agri-biotech Applications (ISAAA), Manila
USDA (2005) 'Roles of US agencies', available at http://usbiotechreg.nbii.gov/roles.aspa accessed 10 May 2006

USDA/AMS (various years) 'Cotton varieties planted', Agricultural Marketing Service/USDA, Washington DC

USDA/ERS (2005) 'Adoption of genetically engineered crops in the US', available at www.ers.usda.gov/Data/BiotechCrops/, accessed 12 May 2006

4

Europe: Turning Against Agricultural Biotechnology in the Late 1990s

Yves Tiberghien

Introduction

Since the late 1990s, the EU has emerged as the leader of a precautionary regulatory approach to GM crops. Between 1997 and 2003, a series of regulations introduced a rigorous safety and environmental impact assessment process. The EU also introduced the mandatory labelling of all GM crops and products derived from them with the most demanding threshold in the world (0.9 per cent). Furthermore, between 1999 and 2004 the EU applied a de facto moratorium on new authorizations of GM crops for both consumption and production. As a result of these strict regulations, and strong mobilization by groups opposed to GM products, the EU has a markedly different relationship to GM crops than the US, Canada or Argentina. Except for about 50,000 hectares of maize in Spain and some small maize fields in France, Germany and the Czech Republic, these crops are hardly grown anywhere in Europe. While GM soy and maize are used as animal feed, very few products derived from GM crops are found in European supermarkets. And while field trials continued throughout the 1999–2004 period, they experienced a visible decline after their high point in 1995–1997.

This European turn against the more liberal approach taken by the US, Canada and other countries was not predetermined and came as somewhat of a surprise. Until 1995, European countries were part of an international consensus reached within working groups of the Organization for Economic Cooperation and Development (OECD) and embedded in the 1994 WTO

Agreement on the Application of Sanitary and Phytosanitary Measures (SPS). In OECD working groups and through the SPS, Europe accepted the principle of 'substantial equivalence', a concept by which novel food products having the exact same functions as traditional ones should not be regulated differently. It also accepted the principle of science-based assessment.

The European move toward the 'precautionary principle' and the emergence of a global rift between the North American and European approach to GM crops appeared after 1997–1998. Japan has emerged as another key country in the regulatory camp, even though its bureaucracy managed to lessen the trade impact of GM product regulations through loopholes and high thresholds. Other countries converging toward a more precautionary approach have included Republic of Korea, China, India and Switzerland.

The rift between the North American liberal approach and the European precautionary approach moved beyond the realm of competing national regulations and into the global arena with the Cartagena Biosafety negotiations in 1999–2000. Under strong leadership by both developing countries and the European Union (and against strong opposition by the US, Canada and others) the UN Cartagena Protocol on Biosafety (CPB) was signed in Montreal in January 2000 and took effect in September 2003 after being ratified by 50 countries.

The US vs EU battle over GM crops also reached the WTO in May 2003, when the US, Canada and Argentina launched legal action against the EU's de facto moratorium on GM products. The preliminary decision of the deliberating WTO panel was announced in February 2006 and resulted in European defeat. The panel confirmed that the de facto EU moratorium of 1999–2004 on new GM product approvals resulted in undue delay and was thus in violation of WTO commitments, namely the SPS. The report of the panel was the longest in the WTO's history (1050 pages) and both camps braced for a tense period, beginning with a likely EU appeal and subsequent law suits for economic damages.

This chapter reviews and analyses the regulatory moves taken by the EU regarding GM products since 1996 and provides an explanation for this turn of events. The actual record of the EU with respect to crop approvals, field trials, research and trade is also analysed.

Overview: The sequence of policy change from 1980 until 2005

Until 1996, the EU basically followed a relatively permissive and supportive policy. Like Japan and the US, it saw biotechnology as a key industry for its future competitiveness and channelled research funds accordingly. It approved 18 GM products between 1991 and 1998 and served as a major import market for such products. Within the EU, France was particularly supportive of biotechnology and became a key location for foreign investment and test plots in the late 1980s.

Box 4.1 Chronology of EU regulations

1990–1991: First EU regulations on contained use and deliberate release to introduce an approval procedure (requires unanimity to block).

1997: 258/97 – Novel Foods: mandatory labelling; only product-based (final product); substantial equivalence accepted (no labelling when no GM protein in final product); 1 per cent threshold.

1997: Amsterdam treaty: first adoption of precautionary principle, allowing countries to take safeguarding measures and impede free trade.

24 June 1999: Moratorium decided by EU Council made up of group of core countries.

Directive 49-50/2000: extension to flavourings and additives, with 1 per cent threshold (under Austrian push); only product-based (no labelling when GM protein not present in the final product). But one big innovation: regulation introduces 1 per cent minimum threshold for adventitious contamination (i.e. conventional food where GM material accidentally introduced reaches 1 per cent must be labelled as GM).

2001: adoption of Directive 2001/18 on Environmental Release (field tests, crop approvals) – complex dual-level approval must be enacted by all EU member states, but problems in France; introduces 0.9 per cent threshold for labelling (Directorate-General (DG) of Environment in lead).

2003: Regulation 1829-1830/2003: core regulations on labelling and approval of GM products; creation of one-stop policy (Public Health (DG Sanco) in lead).

Clearly, the EU underwent a major sea change between 1996 and 1999, turning against GM products under the combined pressure of a few individual member states (Austria, Greece and Denmark), the European Parliament (EP), NGO mobilization and a shift in public opinion in the context of the overall crisis of EU institutions. The leadership position of the EU on GM products emerged as a potential rallying point for regaining the trust of citizens in its institutions and projecting a common identity abroad. The key steps in the EU regulatory sequence are presented in Box 4.1.

Economic interests of stakeholders in Europe

Who are the key economic actors in the GM product battle in Europe? Can the move against GM products be attributed to protective steps taken by important economic stakeholders?

First, it is important to note that there is a strong biotechnology industry in the European Union and that this industry has received government
support at both national and European levels since the early 1980s. As of
December 2000, it was reported that the 15 EU members plus Switzerland
counted over 2092 independent dedicated biotechnology firms.[2] Some of
the biggest global firms in agriculture biotechnology are based in Europe
(Syngenta, Bayer Life Crop Science and Limagrain, among others). Interestingly, the countries leading the pack are Germany (504 firms), the UK (448),
France (342), Sweden (235) and Switzerland (93). Of these five top countries, at least three (Germany, France and Switzerland) have been leading
the moves restricting agricultural biotechnology, going against the interest of
industry. The fact is that regulatory moves on GM products have occurred
against the interests and opposition of a strong and growing biotech industry
– one that has been deemed strategic by all levels of government.

Farmers would appear the most obvious beneficiaries of restrictions on
foreign-grown GM crops. However their position is not that clear cut. In
nearly all EU countries, farmers have been late to mobilize on the issue,
coming into the anti-GM coalitions a few years after urban-based groups.
Only in countries with very small average-size farms and a high-value added
agriculture, such as Austria and Switzerland, have farmers' unions clearly
positioned themselves against GM crops – and even then, after a long and
protracted process. In the Swiss case, during the long negotiations leading to
the Gen-tech law of 2001, members of parliament with strong ties to farmers (or part-time farmers themselves) long hesitated to side with socialists and
break their usual coalition with industry and business. Growers initially saw
cheaper GM feed as a good idea that would improve competitiveness and not
affect the quality of meat. In larger countries, such as France, the federation
of maize growers (AGPM) has taken a pro-GM position since the mid-1990s.
Given the production surplus and export situation of France, maize growers are concerned about losing out in a competitive battle over technology.
Given the possibility that the next generations of GM crops might become
acceptable to consumers, these farmers cannot afford to miss the boat of this
new farming revolution, especially in a country that has historically been
at the forefront of agricultural innovations in Europe. In France, the peak
farming organization, representing about 80 per cent of farmers, the Federation Nationale des Syndicats d'Exploitants Agricoles (FNSEA), has taken a
neutral and prudent position, although it did not lobby against GM crops.
In contrast, the union representing small farmers, Confédération Paysanne,
became the leader of the anti-GM coalition in France after 1999. The GM
crops issue allowed the Confederation to move from a minority position to
the centre of public debate and to build unprecedented direct links with urban voters and consumers.

Most economic actors in the food processing industry moved against
GM crops (although at different speeds), but only after being pressured by

consumer groups, whose efforts trickled up from supermarkets to first stage processors. Initially, most were neutral, given that GM products did not affect them directly. Alternatively, the heavy labelling and traceability regulations did bring important burdens and costs to all intermediary actors.

The main anti-GM actors have been environmental and health NGOs (the first movers in 1996), and anti-globalization NGOs (after 1997–1998). Where green parties existed (as in Germany, Austria, Switzerland, France and the European Parliament) they became core political vectors for the opposition. To some degree, consumer associations have lent a voice to the coalition. Unlike in Japan or the Republic of Korea, however, consumer associations were not in the lead. These actors were successful in capturing media attention and shifting public opinion against GM crops, despite initially pro-biotech politicians and bureaucrats.

Thus, the main drivers of anti-GM crop regulatory positions in Europe have not been economic actors. Rather, the key lies in a dramatic shift in public opinion. Based on data collected between 1 November and 15 December, 1999, the 2000 Eurobarometer reveals that Europeans became increasingly opposed to GM foods. Using an EU average (weighted by population), the data show that support for GM crops among those members of the public who had made a decision dropped by 13 per cent between 1996 and 1999. It is worth mentioning that the Eurobarometer's concentration on only that segment of the public leads to a higher level of support in absolute terms. For example, the EU average opposition of 51 per cent (and average support of 49 per cent) in 1999 compares to an average 71 per cent positive response to the statement, 'I do not want this type of food' among the general population.[3] In relative terms, the survey shows that opposition to GM food increased in all 15 EU countries. Although the support rate decreased by only 3 per cent in The Netherlands (78 to 75 per cent) and by 7 per cent in Germany (from 56 to 49 per cent), it decreased by as much as 19 per cent in France (54 to 35 per cent) and 20 per cent in the UK (67 to 47 per cent). In 1999, only five countries showed opposition to GM crops below 50 per cent: Ireland, The Netherlands, Finland, Portugal and Spain.

An EU Commission-financed analysis of the 2000 data by the objective International Research Group on Biotechnology and the Public (Gaskell, 2000) commented, '[in six key states: Belgium, Greece, Italy, France, Luxembourg, and the UK] patterns of widespread public ambivalence about GM foods in 1996 appear to have given way to widespread public hostility in 1999'. The survey further revealed that even supporters of GM foods and animal cloning believed with an outright majority that the two applications 'threaten the natural order' and are 'fundamentally unnatural' (Gaskell, 2000). This shift in public opinion forced the hand of governments and politicians in moving to regulate GM crops.[4]

Support for R&D in biotechnology since the early 1980s and intellectual property rights protection

Research and development

Both the European Union and national governments identified biotechnology and life sciences as core priorities as early as in the 1980s. The EU reconfirmed this priority in 2002 through its major policy initiative, the EU Strategy for Life Sciences and Biotechnology.[5] This programme provides about Euro 770 million (approximately US$1 billion) annually in public funds for life sciences, allocated through competitive calls inviting collaborative R&D projects.

In agricultural biotechnology, field trials have taken place since the late 1980s throughout European countries, run by both private and public institutions. Table 4.1 gives a thorough account of field trials since the adoption of regulation 90/220 in the EU. (Field trials before that were managed at the national level and records are harder to track down.) The table aggregates specific events (not actual fields) and shows the extent of field trials, which peaked in 1996–1999. France, Spain, the UK, Italy and Germany led by numbers of trials.

A recent report by the EU Commission on the progress of the Life Science strategy deplored the negative impact of public opinion and strict regulations on biotechnology research in Europe, saying that 'the lack of progress on the authorizations of new GMOs is having a direct impact on research activities on GMOs and GMO field trials in Europe'.[6] In particular, the following data were quoted:

> A survey among private companies and research institutes active in the field of GMOs was conducted in order to get an overview of basic and applied research activities on GMOs in Europe. The survey reveals that 39 per cent of the respondents have cancelled R&D projects on GMOs over the last four years, giving as the main reasons the unclear regulatory framework and uncertain market situation. The tendency to cancel R&D projects is low for the public sector (23 per cent of respondents have cancelled projects) and higher in the private sector (61 per cent of respondents having cancelled projects).
>
> Furthermore, the number of notifications for GMO field trials in the EU increased rapidly from 1991 to 1998, and declined sharply thereafter (76 per cent decrease by end 2001). In 2001, the Commission's Joint Research Centre (JRC) database, which keeps a register of EU field trials, received just 61 notifications for field trials with GM plants, compared with over 250 in 1998. Such a marked decrease in the number of GMO field trials has not taken place to this extent outside Europe (e.g. in the US). According to the study, there is a response effect to the lack of progress in new commercial release of GMOs as well as the widespread tendency of the European public to reject GMOs.

Table 4.1 *Record of GM field trials in the EU since 1992 (covered by EU legislation on Environmental Release)*

Country / Year	1991	1992	1993	1994	1995	1996	1997	1998	1999	2000	2001	2002	2003	2004	2005	2006	Total
France		1	35	57	69	91	72	70	64	34	17	3	17	11	14	16	571
Spain			3	10	11	16	44	39	39	19	19	17	40	20	24	39	325
Italy			5	19	43	50	46	43	51	18	5	9	2	4			295
United Kingdom		16	17	23	37	27	25	22	13	25	12	5	8	1			231
Germany		3	1	8	12	17	20	18	23	7	8	7	9	10	7	9	159
Netherlands	4	15	9	25	16	10	14	19	5		19	4	4	7	5		156
Belgium		26	16	17	11	7	7	6	8	16	5	8	1	2			130
Sweden					8	10	9	8	19	6	2	2	1	14	2	4	85
Denmark		5	1	5	4	5	10	4	5	1				1	1		41
Finland					1	3	6	3	3	3	1				1		22
Portugal			2	2	1		3	3	1						4	5	21
Greece						1	5	7	6								19
Hungary															10	7	17
Ireland							2	2				1				1	6
Czech Republic															2	3	5
Poland														1	2	1	4
Austria						2	1										3
Iceland														1			1
Norway									1								1
Total	4	66	89	166	213	239	264	244	238	129	88	56	82	72	72	85	2092

Note: Field trials approved under Regulation 90/220/EEC run until 2004. There is some overlap with those authorized from 2002 onwards by Directive 2001/18/EEC.

Source: Joint Research Centre of the European Union, http://biotech.jrc.it/deliberate/dbcountries.asp

Intellectual property rights

Europe has a unified patent law and a central European Patent Office (EPO) based in Munich. However, the EPO is a non-EU body with a larger membership than the EU. The EU has implemented the requirements of the TRIPS agreement and has a well-developed patent law. (Under the European patent law revised in 1998, patentable subjects include proteins, genes, cells, plants and animals, but *not* plant or animal varieties or human organs (Gold, 2001)). In addition, like Japan and Republic of Korea, the EU has significant patent carve-outs – i.e. the authorities retain the authority to withhold a patent over an invention that is deemed to breach either public order or morality (but not public health). In other words, unlike the US and Canada, the EU represents the majority opinion that public order issues ought to be taken into account in the granting of patents. This debate is summarized by Gold (2001) in the following terms:

> The majority opinion among developed countries is that inventions whose exploitation would injure public order and morality ought not to be patentable. Canada, along with the United States, follows the minority view that patent law and morality ought to be kept separate. These countries hold that moral concerns have no place within patent law itself but ought to be addressed specifically, for example, through regulation or in the criminal law.

Nonetheless, the EU Commission has identified the need to implement a Community Patent Regulation, so as to solve differences that remain between member states and to show EU leadership (Commission of the European Communities, 2003).

In March 2005, the EPO became the first patent office to revoke a patent for reasons of prior public use and lack of invention (the so called case of 'biopiracy'). The patent covered products derived from the traditional Neem tree of India. The legal battle had lasted ten years. This position set the EPO apart from its US and Japanese equivalents, seeming to indicate a possibly greater sensibility to issues of traditional knowledge. If followed by other cases or by an increase in litigation, it may fuel concerns about bio patents in the EU.

Record of crop approval since early 1990s (including after the 1998–2004 moratorium)

The import and consumption of GM crop products in the EU has been authorized since 1996. Between 1996 and 1999, 12 individual events were authorized for consumption (before the de facto moratorium took place). Table 4.2 shows the details of authorizations in the EU until 2006 under various regulations.

Table 4.2 *Record of GM crops approved in the EU for food and feed since 1991*

	1991	1992	1993	1994	1995	1996	1997	1998	1999	2000	2001	2002	2003	2004	2005	2006	TOTAL
Maize							1[7]	4[8]						2[9]	2[10]	1[11]	10
Soybean						1[12]											1
Oilseed rape						1[13]	3[14]		2[15]	1[16]							7
Cotton												2[17]					2
Bacillus subtilis										1							1
TOTAL						2	4	4	2	2		2		2	2	1	21

Notes: All products authorized under Regulation 258/97 (for food), Directives 2001/18/EC and 90/220 (feed and release into environment) and Regulation 1829/2003 (feed *and* food). Most events were authorized through two EU legislative documents. If a product was authorized through one regulation in 1997 and another in 1998, only the first authorization is counted, however, the footnotes indicate subsequent authorizations.

Source: European Commission, http://europa.eu.int/comm/food/food/biotechnology/authorisation/index_en.htm

Regarding production, to date, six events have been approved for production in the EU:

*   Under Directives 90/220/EEC and 2001/18/EC:
        *Bt*176 maize – 23 January 1997
        MS1, RF2 swede rape – 6 June 1997
        T25 maize – 22 April 1998
        MON 810 – 22 April 1998
*   Under Regulation 1829/2003:
        *Bt*11 maize – 19 May 2004
        DAS1507 maize – 3 March 2006

On this basis, several types of maize are currently grown in Spain, France, Germany and the Czech Republic. No soybean has even been authorized for production in the EU to date.

Regulation: Environmental assessment, labelling, Cartagena protocol and trade impacts

Sequence of regulatory moves regarding safety assessments and environmental assessments (1991 regulation, moratorium after 1998, new post-2004 system)

The first major step in the EU regulation of GM safety came with two regulations in 1990–1991. Even though the DG XI (Environment) had gained leadership over GM crops within the Commission by the late 1980s and led the way to the drafting of two restrictive regulations in 1990 – these two regulations made it difficult for the EU to block the approval of GM products. Only in June 1999 did the EU Council reach a broad agreement in favour of a de facto moratorium on new approvals of GM crops. The moratorium was the result of a declaration by a majority coalition of states in the context of a meeting of the EU Council on 24 June 24 1999 (without the grudging acquiescence of the other states). Yet it was not a formal legislative act.

After several intermediary steps, the key regulations for the safety approval of GM products – Regulations 1829/2003 (food) and 1830/2003 (feed) – came in 2003. These regulations create a one-stop approval process for both food safety and environmental release managed by DG Sanco (Public Health). Producers must apply in one member state, which passes the application to DG Sanco. In turn, the newly created, independent European Food Safety Agency (EFSA), based in Parma, Italy, does the technical risk assessment. EFSA bases its assessment strictly on scientific considerations and does not formally consult with social and economic actors. The actual evaluation is partly subcontracted to national evaluation agencies. On the basis of EFSA's recommendation, member states then vote on approval of the new product. If they do not approve the recommendation, they can only block it with a qualified majority of about two thirds. If they fail to block approval with this level of support, the European Commission takes charge and approves the authorization based on a vote among Commissioners. The several cases approved by the EU in 2004 and 2005 all followed this circuitous route.

Environmental impact assessment: Regulations and politics

The key directive in place to regulate environmental release is Directive 2001/18, which to date has been 'transposed'[18] by all 25 members, except France, which is undergoing a long parliamentary process on the directive following a full-fledged parliamentary inquiry in 2004–2005 and should complete transposition in June 2006. The Upper House (Senate) passed the bill on 24 March 2006. Directive 2001/18 replaced the earlier directive passed in 1990 on environmental release, tightening the process. Regulation 1829/2003 now allows a combined safety and environmental review, although few (if any) approvals have yet gone this route to date.

All new GM crops must undergo environmental assessment before commercial release and before field trials. The complex system requires national approval in one country first, based on national laws, and then EU approval as a second step. The lack of specifications on coexistence and liability in the current directive remains controversial, and gives rise to further NGO mobilization.

Labelling

The overall picture is as follows:

- Mandatory Labelling since 1997 (partial) and 2003 (full system);
- Labelling Focus: Process-based (whether or not new DNA present in product);
- Mandatory Labelling of Feed: Yes;
- Mandatory Traceability;
- Threshold = 0.9 per cent.

The EU labelling system is the toughest in the world, combining both a very rigorous threshold (0.9 per cent, and possibly as low as 0.1 per cent in cases where the presence of GM contents is known *ex ante*) with a thorough process of traceability from farm to fork. Labelling is triggered by a GM technology process at any point in production, not by the presence of modified DNA in the final product.

The first labelling requirement came in 1997 with the Novel Food regulation. This regulation, however, was only product-based and had a 1 per cent threshold for adventitious presence. It did not cover all products. The same situation prevailed in 2000 with the regulation on additives and flavourings. To this day, additives and enzymes are not labelled if modified proteins are not present in the final product.

The tougher labelling conditions were introduced with the 2001/18 directive and the 2003 regulation. The impact of labelling has been the search for alternative ingredients, as retailers and producers shun the actual use of labels (given black lists managed by Greenpeace). However, there are reports that consumers may be getting ready to buy some labelled GM products, beginning with a Swedish beer using GM maize. The EU is leading several surveys on the matter.

Impact on trade policy and trade flows

Trade flows between Europe and North America have been affected by the slow process of GM approval in Europe. Maize imports from the US collapsed (from US$400 million in 1996 to US$15 million in 1999 and US$10 million in 2004)[19] and soybean imports decreased significantly (US$2.3 billion in 1996 or 1997; US$1.0 billion in 1999 and US$0.9 billion in 2004). Imports were mostly redirected toward Brazil and other GM-free producers, although Brazil's shift to GM products after 2003 raises questions about the sustainability of this change. The EU may yet return to imports from the US in the near future.

It is hard to believe that the EU has gained from this redirection of trade flows. As the price of feed has increased, the impact on the meat industry may have been negative. On maize as well, the main impact of the EU moratorium and EU regulations has been to redirect imports from the US to Argentina and Brazil. In 1995, before the introduction of GM crops, 62 per cent of EU maize imports were internal (from other EU countries), 32 per cent from the US, 5 per cent from Argentina and 0 per cent from Brazil.[20] By 2004, the portion of EU internal imports had modestly increased to 68 per cent, the US share had collapsed to only 1 per cent, while Argentina took 12 per cent and Brazil 13 per cent (with Romania also gaining a 2 per cent share). Regarding canola, the EU was 100 per cent self-sufficient in 1995. By 2004, the US had actually gained a 2 per cent import share but Canada stayed at 0 per cent. Other beneficiaries were Belarus (2 per cent) and Russia (1 per cent), with the EU internal share at 94 per cent. Far more importantly, European regulations have had a long-lasting negative impact on the EU biotech industry, undermining its potential technological leadership for decades to come.

It is often alleged that protectionism drove the EU move toward a moratorium on GM products. Anderson and Jackson (2006) argue that European industry is lagging in biotech research and thus advantaged by any slowdown. They further contend that European farmers welcome the ban, given their structural inadaptability to GM crop farming (due to small farm size and difficulties in implementing large buffer zones). However, analysis of the empirical record shows that both groups strongly opposed the moratorium, supported GM crops on the whole and did not originate the decisions against them. Europe's biotech industry has close links with its US counterparts and is integrated into an effective EU-level lobbying organization, Europabio. Companies such as Aventis, Bayer Life Crop Sciences and Syngenta have made strong investments in biotechnology and see it as a priority. They would stand to gain significantly if the industry took off. As for farmers, all significant large farmers' organizations, beginning with European Consensus – Platform for Alternatives (ECOPA; the EU peak organization), have supported GM crops and still do. For large farming lobby groups in France, Germany and Spain or at the EU level, it is simply suicidal to lag behind a core technology that is transforming agriculture. At a recent EU summit on GM products in April 2006, possibly the most visible clash pitted pro-GM farmers against anti-GM NGOs such as Friends of the Earth and Greenpeace.

Role in Cartagena Protocol for Biosafety

The EU was the chief proponent of the CPB, in coalition with developing countries. The EU saw the CPB as a way to simultaneously cement its regulatory leadership, increase institutional legitimacy in the eyes of EU public opinion and build stronger linkages with developing countries. EU countries quickly ratified the CPB in 2002 and passed legislation to implement it.

A political explanation of the move towards strong precautionary policies

The EU began to develop a unified regulatory response to biotechnology in 1990 with two core regulations on approval mechanisms for GM products. Until then, member states were the prime movers and took distinct approaches. With the 1990 regulation, the EU decided to treat biotechnology products separately from other products and developed a specific approval procedure. However, this approach remained compatible with a science-based attitude and did not lead to major differences with the US and other partners. The key changes came in 1997–1999, spurred on by several political factors.

Three member states acted consistently against GM crops throughout the 1990s: Denmark, Greece and Austria. Denmark initiated the EU regulatory process through its own 1986 Environment and Gene Technology Act, while Greece and Austria consistently acted as first movers and initiators of EU-level anti-GM regulations during the 1990s. In 1997, an Austrian public petition against GM crops garnered signatures from 20 per cent of voters. By contrast, the big states, particularly France, Germany and the UK, wavered back and forth. Steeped in its role as a European agricultural powerhouse and its strong presence in the biotech industry, France took a position as the sole pro-GM state in 1997 and 1998, allowing the EU to approve the notorious Novartis maize, *Bt*11. But by 1999, France had turned resolutely anti-GM, leading the battle against the US in words and at the WTO.

The position taken by each country is the result of a combination of interests and embedded values. The Austrian position seems rooted in a rejection of eugenics, strong environmental awareness and a refusal to accept the domination of the EU by big countries. The Greek position seems more ad hoc, heavily influenced by a successful Greenpeace campaign in the country before public opinion was formed. In France, by contrast, strong industrial and agricultural interests have formed a de facto pro-GM elite position, still present to this day. The national position was forced to change by the (unexpected) grassroots mobilization of civil society groups and strong public opinion response. The UK's situation is similar (without the strong agricultural interests).

However, the shift in the EU position cannot be explained by inter-state politics alone. A key element seems to have been effective campaigns by NGOs and civil society groups that may have thrived upon the existing public malaise over the institutional capacity of the EU in the context of globalization.

Who were the key civil society groups involved and what motivated them? Unlike in Japan, where consumer groups were at the forefront, the first movers in Europe were environmental groups, such as Greenpeace and Friends of the Earth. Also active since the late 1980s were influential green party leaders from Germany and other countries, such as Benedikt Haerlin and Friedrich-Wilhelm Graef zu Baringdorf. Both Haerlin and Baringdorf were members of the European Parliament. Within Greenpeace, there was initially a sharp debate, as its global headquarters opposed the idea of a campaign against GM products. A key discussion pitted French leader Arnaud Apoteker against the international leadership. Only in 1996 was the anti-GM campaign approved. Apoteker and Greenpeace were then stunned by the overwhelming impact of their campaign.[21] For European environmental NGOs, GM crops became the symbol of a production-focused agriculture that would destroy the environment. They saw GM crops primarily as tools to increase productivity above all else and questioned the incentives of the biotech industry. Yet they quickly realized that adding a health dimension to the issue would broaden its appeal. In the initial 1996 campaign, health and food safety dominated. In a second step, the anti-GM civil society groups became a larger coalition that included minority farmers' groups, anti-globalization NGOs, consumer groups and some religious organizations (though not to the same extent as in Republic of Korea). GM products became the visible face of anti-globalization resistance in Europe. This new focus after 1999 captured public opinion in countries like France and forced the hand of the EU Council and Commission in moving against GM products. A final dimension was the transitional situation of the European Union, at the time caught in the midst of institutional change and a fierce debate on democratic deficit. Initial moves by the Commission in favour of GM products in the context of fearful public opinion came to be seen as a prime example of the lack of the EU's democratic accountability. This made it nearly impossible for the EU to retain its benevolent policy toward GM products.

Civil society groups found support in a European Parliament willing to enlarge its influence. The EP intervened at key junctures, particularly after 1996. In November of that year, at the height of the Greenpeace campaign against the delivery of GM soy from US ships, the EP passed a resolution calling for compulsory labelling, thus lending political support to the Greenpeace campaign. In 1997, the EP condemned the Commission's certification of Novartis maize. It later called for an EU moratorium. In turn, the Council of Ministers (representing member states) and the Commission responded to these bottom-up pressures with a strong regulatory push. The Council coalesced around the de facto moratorium in June 1999, while the Commission, energized by new and enlarged DGs for Environment (enlarged in 1992) and Health (created in 1992) moved toward regulatory leadership.

However, the situation in 2004–2005 showed that the tight EU regulations might have been partly successful in rebuilding trust in a few countries (UK, Denmark, The Netherlands), but not in many others (France, Germany, Greece, Italy). GM products continue to attract considerable political attention in many countries and many regions. With the EU and some states (including France) attempting to promote some level of research, the battle with civil society groups continues and the situation remains unstable.

One of the most recent political developments is the rise of a network of GM-free regions. The movement started in 2004 with Tuscany and Upper Austria, but soon snowballed to include 36 regions (mostly in Greece, Italy, Austria, France, Scotland, Wales and Spain) and nearly 3200 local government units. The regions are relying on the leadership of some countries (particularly Austria and Italy), of the green parties and of the EP to advance a more radical anti-GM agenda.[22] In the latest twist, the Swiss referendum of 27 November (in which 55 per cent of Swiss voters and all 26 cantons approved a five-year moratorium on GMO production) emboldened Austria, the holder of the EU presidency from January 2006 to call for a 4–5 April EU summit on GM products in Vienna (see below).

Future perspectives: Towards new regulations on coexistence?

The GM crop scene remains fluid and unsettled in the EU. On one side, the European Commission, some member states and biotech industrial groups are trying to use the background of thorough regulations to push research and production forward. The hope is that the current regulations can bring higher public acceptance and that progress in biotechnology can rebuild enthusiasm for the technology. Several countries have seen improvements in public opinion toward GM crops and are increasingly supporting approvals of new products.

A strong constituency favouring agricultural biotechnology remains. In addition, the outcome of the WTO panel on GM products – against the EU – confers a legal basis for the Commission to loosen up its attitude. Although EU regulations were not seen as illegal per se, the de facto moratorium on new approvals enforced by the EU in 1998–2004 was seen as illegal according to WTO SPS rules. This outcome delivered an important victory to the pro-GM camp – one that is likely to reverberate in other countries and could lead to a new case against the EU, this time challenging the core of its regulations. At the same time, this could also be a pyrrhic victory, as it puts wind in the sails of anti-GM activists and weakens the legitimacy of the WTO in the eyes of European public opinion.

Meanwhile, the Swiss referendum of 27 November 2005 (with its 55 per cent in favour of a five-year constitutional moratorium on GM crops) has renewed the pressure on the EU to move further and to pass strict regulations on coexistence. The April 2006 summit in Vienna called for by Austria brought over 700 government officials and stakeholders into the largest open EU-sponsored debate, confirming deep splits among stakeholders. It gave rise to strong calls for further EU regulations on seed purity (with a 0.1 per cent threshold appearing as one potential target) and for coexistence rules between GM crops and non-GM crops, particularly to preserve pure organic agriculture.

Notes

1   The author wishes to acknowledge the generous funding of the Social Science and Humanities Research Council of Canada and the able research assistantship provided by Marko Papic for this chapter, which is part of a larger research project on the global politics of GM products. More information is provided at www.gmopolitics.com. The Chapter draws upon Tiberghien, 2005 and 2006.

2   Source: Pamolli and Ricaboni, 2001, University of Sienna data, quoted by Euroapabio at www.europabio.org/images/DBFS-HR.jpg.

3   Eurobarometer 55.2 conducted in 2000.

4   For excellent analyses on the shift in public opinion on GM products in Europe, see also Gaskell (2000); Gaskell et al (2001); Bauer and Gaskell (2002); and Bonny (2003).

5   Source: www.europa.eu.int/comm/biotechnology

6   Commission of the European Communities, 2003.

7   *Bt*176, approved under both 90/220 and 258/97 in January 1997.

8   MON 810, T25 and *Bt*11 (Novartis variety) received approvals under 258/97 in February 1998 and were also approved under 90/220 in April 1998. MON 809 was approved only for food use under Novel Foods Regulation 258/97 in October 1998.

9   NK 603 authorized under 2001/18 in July 2004 and also authorized under

1829/2003 in October 2004. *Bt*11 (Syngenta variety) only authorized under 1829/2003 in May 2004.

10 MON 863 first authorized under 2001/18/EC in August 2005 and then under Regulation 1829/2003 in January 2006 (for food); also DAS1507 maize, first authorized under 2001/18/EC in November 2005 (for feed and industrial use) and then authorized under Regulation 1829/2003 in March 2006 (for food).

11 GA21 only authorized under 1829/2003 in January 2006.

12 Soybean 40-3-2 only authorized for import and processing (under 90/220) and for food uses (under 258/97), both in April 1996. However it is authorized for cultivation in Romania, where it is widely grown.

13 MS1/RF1: in February 1996 only authorized under 90/220 for 'breeding purposes'; in June 1997 also authorized under 258/97 for food.

14 TOPAS 19/2, MS1/RF2 and GT73 all authorized for food use under 258/97. Of these, MS1/RF2 also authorized in 1997 under 90/220. TOPAS 19/2 finally authorized under 90/220 in April 1998 and GT73 finally authorized under 2001/18 in August 2005.

15 Falcon GS40/90 and Liberator L62 only authorized for food under 258/97 on November 1999.

16 MS8/RF3 only authorized for food under 258/97 in April 2000.

17 Both cotton varieties 1445 and 531 only authorized under 258/97 (for use in food) on December 2002.

18 In the EU policy-making process, EU directives are general framework bills that still need to be adapted and passed by each national parliament. This process of adaptation and incorporation into national law by national parliaments is referred to as 'transposition'. According to EU treaties, countries have a fixed amount of time in which to complete the process and the Commission can sue states that are late in the European Court of Justice. The Commission launched this legal process in the case of France and the GMO directive in 2006.

19 Source: United States Department of Agriculture, Foreign Agriculture Service.

20 Percentages calculated on total quantities, using statistics from the United Nations Statistics Division: http://unstats.un.org/unsd/comtrade/dqBasicQuery.aspx

21 Source: personal interview with Arnaud Apoteker, June 2005.

22 See the report of the latest meeting of regions in Bretagne, November 2005: http://www.region-bretagne.fr/CRB/Public/toute_lactualite/les_archives/economie/4e_conference_du_res_11327583859508

References

Anderson, K. and Jackson, L. A. (2006) 'Transgenic crops, EU precaution, and developing countries', *International Journal of Technology and Globalization*, vol 2 (1/2), pp65–80

Bauer, M. W. and Gaskell, G. (2002) *Biotechnology: The Making of a Global Controversy*, Cambridge University Press in association with the Science Museum, Cambridge, New York and London

Bonny, S. (2003) 'Why are most Europeans opposed to GMOs? Factors explaining rejection in France and Europe', *Electronic Journal of Biotechnology*, vol 6 (1), pp7–8

Commission of the European Communities (2003) 'Communication from the Commission to the European Parliament, to the Council, and to the European Economic and Social Committee: Life sciences and biotechnology – a strategy for Europe – progress and future orientations', EU Commission, Brussels

Gaskell, G. (2000) 'Biotechnology and the European public', *Nature Biotechnology*, vol 18, pp935–938

Gaskell, G., Bauer, M. W. and Science Museum (Great Britain) (2001) *Biotechnology, 1996–2000: The Years of Controversy*, Science Museum, London

Gold, R. (2001) 'Patenting life forms: An international comparison', Canadian Biotechnology Advisory Committee, Ottawa

Tiberghien, Y. (2005) 'Agriculture biotechnology policy-making in the EU, Japan, Korea, Chinese Taipei, and Australia: The balance between national politics and global trade commitments', *Building Capacity on Trade and Biotechnology Policymaking*, International Centre for Trade and Sustainable Development (ICTSD), Geneva

Tiberghien, Y. (2006) 'The battle for the global governance of genetically-modified organisms: The roles of the European Union, Japan, Korea and China in a comparative context', *Les Etudes du CERI*, Institut des Etudes Politiques (Sciences Po), Paris

5

West and Central Africa: Strategizing Biotechnology for Food Security and Poverty Reduction

Marcel Nwalozie, Paco Sereme,
Harold Roy-Macauley and Walter Alhassan

Introduction

The West and Central African subregion, comprising 22 countries, is home to some of the world's poorest people. Among its population of 280 million, more than 120 million live below the poverty line of US$1 a day as defined by the World Bank (2000, 2003).

Agriculture is the most important sector of the WCA economy, employing some 70 per cent of the population. It contributes about 15.3 per cent of total export earnings (30 per cent if Nigeria is excluded), as well as from 35–60 per cent of the subregion's total GDP (2000, 2003). Yet the agricultural potential of the subregion remains largely under-exploited. Yields and productivity per farmer are among the lowest in the world and food accounts for about 19 per cent of all imports. Inadequate investment in agricultural research has contributed to this poor performance. Since independence, low agricultural productivity has been a main source of the subregion's persistent problems of hunger, poverty and malnutrition, as well as its disappointing economic growth (see Figures 5.1 and 5.2).

The most important challenge facing the subregion is to triple current food availability by 2025. Given the imperative of producing more food from the same (or less) land while boosting nutritional value and reducing environmental impact, the application of modern biotechnology could be of special relevance.

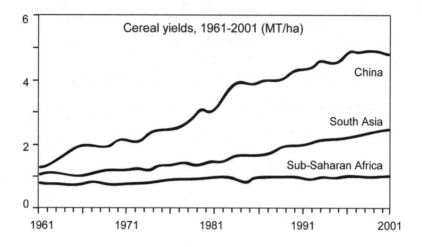

Figure 5.1 *Comparing cereal yields by region, 1961–2001*

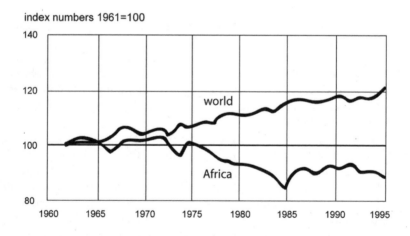

Figure 5.2 *Comparing per capita food production trends: Africa vs rest of the world*

Biotechnology in the context of national and subregional priorities

The potential of biotechnology in meeting national priorities

As described in Chapter 1, three basic biotechnology tools can be applied to improve conventional crop and animal production: tissue culture, molecular breeding and genetic modification. Of these, tissue culture is the simplest and least expensive, while genetic modification is the most sophisticated, requiring highly trained scientists and costly infrastructure. These techniques could complement conventional approaches to developing crop varieties in the subregion.

Tissue culture is important because it produces plants that are generally stronger, reach maturity earlier, are free of pests and most diseases, and have higher and better quality yields. The use of molecular breeding in programmes in the subregion is expected to be particularly useful in tackling food crops' adaptation to drought and increased resistance to pests and diseases, responsible for the lion's share of losses. Genetic modification can produce plants or animal types that might otherwise have been developed with great difficulty – or not at all.

In WCA, the application of GM technology has been limited to Burkina Faso, where controlled field trials on *Bt* cotton are being undertaken in collaboration with Monsanto. The majority of research carried out on roots, tubers and tree crops is on mass propagation through tissue culture. Few national laboratories are endowed with the capacity to carry out germplasm characterization and molecular diagnostics on these crops. In the case of cereals, too, very few national laboratories do germplasm molecular characterization and fewer still have the capacity for using molecular markers to assist breeding and selection. To bridge this biotechnology research gap, there is an urgent need to: first, gradually assist laboratories at the tissue culture stage; second, move into germplasm characterization and specific molecular marking technologies; and third, identify a desirable gene or groups of genes to solve the intractable problems of resistance to biotic and abiotic stresses.

Important staple food crops in the subregion with serious disease and pest problems include cowpea, maize, sorghum, cassava, cocoyam, cocoa and coconut. Cotton is similarly affected. To improve resistance – and consequently production – research using conventional approaches is currently underway. However, these crops could benefit from genetic transformation and other biotechnology research tools.

To overcome productivity and disease limitations in animals, countries need to develop research capacity in diagnostics, molecular characterization of livestock and poultry and its linkage with breed improvement, and ability to develop and test recombinant vaccines against diseases such as contagious bovine pleuropneumonia (CBPP) and heartwater (cowdriosis).

The constraints

A study has revealed important constraints to the wider application of biotechnology in the subregion, including weak research capacity, little political commitment, poor private sector linkage and controversies surrounding GM crops related to a lack of information (Alhassan, 2003).

National agricultural research services (NARS) have been categorized as three types according to their capacity: Type I, with molecular biology and plant breeding capacity; Type II, with limited molecular biology but solid plant breeding capacity; and Type III, with limited capacity overall (Byerlee and Fischer, 2002). All the NARS in WCA fall within Type III. They have plant breeding capacity, but small fragile programmes usually depending on one or two individuals. Human, financial and infrastructure resources to implement promising lines of research beyond the pilot scale are limited. Partnerships with the private sector for food crops are almost nonexistent. As modern biotechnology requires advanced scientific research capacity, successful use of agricultural biotechnology in WCA will require increased capacity and supporting infrastructure, including working links between the scientific community and end users.

Another major constraint is the regulatory structure. Most WCA countries have biosafety rules and regulations but these concern phytosanitary issues and the import and export of seeds. While all countries are currently taking further action in this area, some have yet to establish a formal position on GM crops. In 2001, the Organization of African Unity (now African Union) proposed a model law based on the Cartagena Protocol on Biosafety. Cameroon has already ratified this protocol; all other countries in the subregion have signed it and are in the process of ratification.

Lack of political commitment and public acceptance are other barriers. There is a small but vocal opposition to introducing new technology in WCA. Concerns focus on risks to the environment and human health but also include: multinationals' lack of interest in the region's traditional food crops; farmers becoming dependent on purchasing seeds from multinational suppliers rather than saving their own seeds; and inadequacy of regulatory structures to assure environmental stewardship and prevent seeds from crossing borders. African policy makers appear to be taking a 'wait and see' attitude. However, this scepticism is often based on meagre scientific information that has had little dissemination, while much unscientific data has been circulated on potential risks of GM products. Negative perceptions are often based on local decision takers' concern for certain donors' positions rather than scientific analysis.

A major challenge is to disseminate information on the potential benefits of biotechnology for food security, putting in perspective the controversies over the role of multinationals reaping huge financial benefits by selling to the wretched of the Earth and not allowing farmers to save their own seeds.

Embracing the new initiative: Engaging the region

The decision to develop a regional biotechnology programme under the Economic Community of West African States (ECOWAS) grew from a series of international and regional meetings in 2003 and 2004. Subsequently the ECOWAS ministers of agriculture adopted recommendations that provide a strategic basis for this programme, including:

• encouraging and increasing investments in agricultural biotechnology through stimulation of private-public partnerships, and encouraging networking of national laboratories engaged in biotechnology and mobilizing the African Diaspora;
• promoting the establishment of a regional policy and national systems on IPR management in order to facilitate the acquisition, development and dissemination of knowledge and innovation;
• establishing a regional regulatory framework and strengthening national seed production systems to facilitate the distribution of improved seeds, and encouraging all countries to ratify the Cartagena protocol by June 2006 to facilitate harmonization across the region by the end of July 2008;
• putting in place an independent fund for assessment of the socio-economic impacts of the use of GM crops;
• supporting countries in creating national information and communication units charged with raising public awareness of the benefits of applying biotechnology to agriculture, and supporting and encouraging ECOWAS and its efforts to establish partnerships with other organizations having experience in this area.

The West and Central African Council for Agricultural Research and Development (CORAF/WECARD) was designated to play a lead role in this initiative, and to act as a technical arm for the implementation of research and development activities. CORAF/WECARD is a subregional institution established in 1987 to promote cooperation among member institutions and its partners.

CORAF/WECARD's approach to agricultural biotechnology research and development in the subregion

An in-depth study (Alhassan, 2003) undertaken by CORAF/WECARD revealed that WCA countries vary greatly in terms of human resources, infrastructure and commitment to the use of biotechnology as a tool to enhance agricultural and general socio-economic development. The study proposed prioritization and management of biotechnology activities in West and Central Africa within the current CORAF/WECARD network management process and identified three laboratories with quantitative *trait loci* mapping capacity that could be used as training grounds while solving subregional problems requiring high-level biotechnology intervention: Centre d'Etude Regional pour l'Amelioration de l'Adaptation a la Sécheresse in Senegal, the

Cocoa Research Institute in Ghana and Centre National de Recherche Agronomique in Côte d'Ivoire. It proposed that these be targeted for initial support.

Since most countries in the subregion cannot afford to go it alone, the regional pooling of facilities and expertise is the only logical way to tap the potentials of this new technology. Transformation of some of the regional/international centres into centres of excellence for biotechnology could provide training facilities for NARS scientists while creating centres for technology generation and adaptation for the subregion.

CORAF/WECARD would work on mobilizing WCA and international experts to carry out biotechnology work in these centres, and on creating multi-disciplinary, multi-institutional and multi-stakeholder partnerships to build biotechnological and biosafety tools and expertise, leading to product development.

The research and development agenda

CORAF/WECARD set up a working group to delineate tasks and propose a subregional position to stakeholders. It comprised experts in biotechnology and biosafety from countries, intergovernmental organizations and international and regional research institutions in the subregion, plus scientific partners and facilitators from the Agricultural Biotechnology Support Project II coordinated by Cornell University, and from the Programme on Biosafety Systems coordinated by the International Food Policy Research Institute (IFPRI).

Priority setting process

Based on the two inter-related priorities of agricultural development in the subregion – achieving food security and reducing poverty – work plans, a budget and an implementation programme were developed.

Priority commodities were selected based on relevant criteria such as economic growth, social welfare, environmental quality, capacity development and potential impact. Further prioritization of crop commodities by each of three zones (Sudano-Sahelian, Wet Sub-tropical West and Wet Sub-tropical Central)(CORAF/WECARD, 2000, 2004) resulted in ranking a long list of priority crops and animal products, followed by identification of a relatively large group of common production constraints (see Table 5.1). An economic analysis of the priority commodities and constraints is presently being conducted by CORAF/WECARD in partnership with IFPRI and will be used by the CORAF Biotechnology and Biosafety Programme (CORAF-BBP) to determine priority targets for the application of biotechnology research.

The process included the identification of opportunities to adapt existing research from the international community. It also focused on near-term results that could be adopted or adapted by WCA countries within two to five years, analysing the near-term impact of various agricultural biotechnology options in terms of their economic, social, environmental, technical, trade and health consequences to help CORAF/WECARD rally support from investors and the political caucus in the subregion. It considered a number of potential research projects/programmes where biotechnology applications are already under consideration.

Working Group recommendations were then adopted by a wider stakeholder forum that included: policy makers within national governments and regional economic communities, managers and scientists of national agricultural research systems and international agricultural research institutions, development partners, NGOs, civil society farmers' organizations, professional agricultural organizations, the private sector, the media and donors.

The programme

The outcome of this whole process was a CORAF-BBP strategy document for the subregion based on:

- demand-driven priority setting;
- a strong focus on product development and delivery involving South–South and North–South collaboration;
- support for activities through a complimentary combination of a competitive grants system and commissioned projects.

The anticipated products range from urgently needed new crop varieties and crop propagation methods to novel diagnostic kits for the detection of animal or plant diseases, new vaccines and reproductive technologies for improved livestock production. This product-driven approach to capacity building will provide 'real life' experience for scientists, regulators, extension workers and farmers. It will also improve the general public's ability to make informed decisions based on benefits and possible risks associated with each product, rather than on generalized debates. The biotechnology products themselves are expected to directly contribute to boosting food and nutrition security, economic growth and environmental quality.

It is important to note that technology solutions – neither biotechnology nor any other type – may not be currently available to address some of the key productivity constraints identified during the priority setting process. Also, the priorities established today may need revision in the future. The process must continue, integrating supporting analytical and quantitative analyses that take into consideration such issues as linkages of commodities and sub-sectors to agricultural development domains, interactions among supply and demand, prices and traded quantities, and linkages between agricultural and non-agricultural issues and the overall economy.

Another key issue is how results can be modified to guide biotechnological interventions while addressing differences among the countries concerned.

This could be realized by targeting investments proportionally to enhance the activities of farmers, the private sector and local governments based on:

- *concordance* (with farmers' resource allocation decisions);
- *catch-up* (weighing probabilities of success in the light of breakthroughs achieved elsewhere);
- *complementarity* (supporting rather than replacing activities of private entrepreneurs and public agencies).

Product development and delivery focus

WCA's CORAF-BBP focuses on product development and delivery for improving farm productivity. Important issues to consider are: whether or not the end-user will have access to the product; how the product will be made available to him or her; and how the end-user would gain access to the information necessary to make a rational choice.

In developing products, near-term technology opportunities will focus on the transfer and use of proven biotechnologies that must be adapted to priority local crops and livestock and their production constraints in the subregion. Longer term technology opportunities will require mobilizing the necessary scientific, human and material capacities before proceeding with certain activities. Private sector actions to stimulate development of seed production and commercialization of biotechnology products will be important here, with sustainability depending on the degree of involvement of private companies, including multinationals, especially in the development of GM varieties. In terms of product delivery the principal challenge will be to adapt and utilize biotechnology as an integral part of the seed multiplication system.

The subregional approach envisions:

- commissioning research on the following target crops/livestock commodities: sorghum, maize, cassava, cattle, goat and sheep, and their production constraints, based on near- and long-term technology opportunities identified (see Table 5.1);
- scaling up investments so that the technologies produced can reach the millions of dispersed, resource poor farmers who need them.

Table 5.1 *Summary of top ranking crops and livestock and the most important problems associated with their production, derived from the CORAF/WECARD priority setting exercise*

Crop	Constraint
Sorghum	Striga resistance
	Insect resistance (bugs, borers etc.)
Maize	Grain protein quality
Rice	RYMV resistance
	Pyriculariose resistance
Groundnut	Aflatoxin control
	Resistance to rosette and clump viruses
	Control of storage insects (weevils)
	Resistance to fungi (rust, Cercosporia)
Cowpea	Resistance to Striga
	Resistance to post-harvest insects (weevils)
	Resistance to insect pests affecting production (bugs, pod borers)
Tomato	Resistance to Tomato geminivirus
	Modified ripening
	Nematode resistance
Coconut	Resistance to lethal yellowing
Banana/plantain	Nematode resistance
	Resistance to viruses (CMV, BTV, BSV)
Cassava	Resistance to the ACMV
Cotton	Resistance to *Bemiscia tabaci* and Helicoverpa
Cacao	Resistance to *Phytophtora* sp

| Forestry crops | Seed production |
| All crops | Maintenance and evaluation of genetic resources |

Livestock	Constraint
Goats	PPR
	CBPP/PPCB
	Maintenance/evaluation of genetic resources
	Heartwater
	Helminthiasis
Sheep	PPR
	Maintenance/Evaluation of genetic resources
	Heartwater
	Helminthiasis
Bovine	Trypanosomiasis
	CBPP/PPCB
	Maintenance/Evaluation of genetic resources
	Tsetse
	Heartwater
	Foot and Mouth Disease
	Helminthiasis
Pork	African Swine Fever
Poultry	Newcastle Disease
	Helminthiasis

Investments required for agricultural biotechnology in West and Central Africa

Public investment

Most agricultural research in Africa is carried out by publicly funded national agricultural research institutions, which generally perform poorly. In building capacity to conduct development-oriented agricultural biotechnological research, there would be some initial phase of apprenticeship with partner institutions, notably the advanced research institutions from the North, the international agricultural research centres of the Consultative Group on International

Agricultural Research (CGIAR), and possibly multinationals. Though it is difficult to quantify financial requirements for capacity development, it may be worth considering CGIAR investments as an example.

Four of the 15 CGIAR centres are Africa-based and two (the International Institute for Tropical Agriculture (IITA) and the West African Rice Development Association) have headquarters in West and Central Africa. Africa alone absorbs half of CGIAR's total annual budget, which in 2003 was US$398 million. In addition to the CGIAR system, there are a number of independent international and regional agricultural research and academic institutions based in Africa that are also involved in capacity building and human capital development, not only in biotechnology, but also in other spheres of agricultural research for development.

Within the next five years, an estimated US$19.3 million of new money is expected to be mobilized from investors for the implementation of public research in biotechnology and biosafety in the subregion. This does not include the probable contributions from the national governments in cash and kind.

Investment from the private sector

In Africa, private research is concentrated on food processing and post-harvest technologies (in 1995, only 12 per cent dealt with farm-focused technologies) and the farm technology needs of commercial agriculture (seeds, vaccines, agrochemicals and farm equipment), where profits can be captured. The private seed and chemical sectors remain very poorly developed in the smaller markets of Africa.

Private agricultural research relevant to African agriculture is likely to develop as profitable markets become available. Meanwhile, African agricultural research will continue to rely overwhelmingly on public support through NARS and international centres – especially activities addressing the needs of the poorest farmers.

Institutional challenges

Seed systems

In WCA two seed systems coexist side by side, the informal and the formal sectors. The informal (traditional) seed sector provides 85 per cent or more of planting seeds among smallholder farmers and involves the collective efforts of farmers and their local communities. Much of the information on seeds concerning agronomic performance, yield, disease resistance, quality, cultural preference and diversity of end uses comes by word of mouth; seldom, if ever, is it subject to rigorous experimental evaluation. Relying largely on local resources and inputs, seed supply in this sector is very vulnerable to disaster and socio-political disruption.

The formal seed sector generally comprises public (national, regional and international agricultural research and policy) organizations and private businesses involved in plant breeding and research on related aspects of seed physiology and plant disease variety release, as well as seed multiplication on seed farms, and seed processing, storage, marketing and distribution. This sector tends to focus on larger farming operations and hybrid varieties.

In the WCA subregion, the informal and formal seed sectors are neither complementary nor interactive. In both, variety testing and regulation and seed quality control are unsatisfactory because of poor organization, inappropriate standards, little or no opportunity for farmer and seed producer involvement and lack of transparency in the seed regulatory process (Maredia and Howard,1998). Reasons include: decreasing national budgets for public sector research; declining donor support for long-term breeding and variety development and control of plant genetic diversity; pressure to establish plant variety rights; emergence of variety development and seed production at the local level; and the collapse of several parastatal seed organizations.

Within the formal seed system, the private sector has focused on species and crops showing high monetary returns. Along with parastatals, private firms have focused mostly on crops of commercial or industrial interest rather than staple crops essential for food security and for the poorest farmers. In general, it is simply not economic for private companies to market self- or open-pollinated varieties.

The informal sector, by contrast, has concentrated on crops and seed systems essential for local food production. This sector is also responsible for ensuring a sustainable supply of propagating material of asexually propagated food crops such as cassava, plantain, yam, potato and sweet potato – an important role often overlooked when seed systems are considered from the perspective of commodity crop production. As small farmers' planting seed is mainly self-stored, or purchased or bartered from local sources, there is an overwhelming dependence on farmer/community-based seed delivery systems to sustain basic food crop production.

Private sector involvement in seed research and development and supply is needed to strengthen partnerships among NARS, international research centres and seed producers. The capacity of public sector departments to work with private enterprises on seed production also needs strengthening. Training should be provided in: norms of control in the field and during seed collection; determination of seed density, varietal purity and rate of fertilizer contamination; packaging and treatment of stored seeds; use of control forms to ensure the availability of sufficient, selected high quality seeds; and intellectual property issues pertaining to seed production, multiplication and distribution. More involvement of the private seed sector in such areas can open up avenues for profitable investments.

But seed policy in WCA member countries also requires management of a security stock to be tapped in times of peril, for which refrigerated seed stores are required. Each country needs a national seed committee to assume responsibility for the homologation of varieties and advise on the whole range of seed issues.

Intellectual property right issues

Practically all the countries of the WCA subregion are members of the WTO and thus obligated to implement minimum standards for IPRs under the Agreement on TRIPs. Several regional institutions in Africa have a direct IPR mandate and/or undertake major activities in this area. Two deal specifically with IPRs, the African Regional Industrial Property Organization and the African Intellectual Property Organization.[1] The African Union has also carried out relevant activities, especially with respect to the interface between IPRs and genetic resources. One of the major challenges for the region in seeking to harmonize rules has been the different approaches and levels of legislative development adopted by countries.

Figuring in almost every aspect of biotechnology development, IPRs must be initially taken into account by all stakeholders (regulators, researchers, private operators and consumers), kept in view through planning processes, and factored into eventual implementation and/or related activities. A key issue is how to acquire, produce and disseminate products and knowledge with the fewest possible constraints while respecting the rights of individuals, organizations and nations recognized by the new international agreements. In this context, institutions and initiatives like the African Agricultural Technology Foundation could play a pivotal and catalysing role in negotiating IPR issues.

Advanced research tools used in agricultural biotechnology tend to be held as intellectual property by an industry or other institution so as to protect its massive investments in research and development. Thus, researchers must know the proprietary status of the technologies and related materials that they intend to access and utilize. IPRs may also be seen as vehicles for technology transfer through licensing or similar agreements. An appropriate IPR policy framework needs to be in place to facilitate effective implementation of a biotechnology programme.

Dual approach to supporting product development and delivery

Financial support for the competitive grants/commissioned research scheme proposed by the West and Central African Biotechnology Programme led by CORAF/WECARD will be conditional on potential grantees forming subregional consortia among and within countries in the subregion and with the international research community, as well as on a pragmatic plan to deliver research products to end-users while building subregional capacity in research and product development.

Both modes of support will be managed through existing CORAF/WECARD decision-making organs. A newly created CORAF/WECARD Biotechnology and Biosafety Coordination Unit, which will include national focal points within NARS, NARS components and centres of excellence within the subregion, will be mandated to implement activities.

Public information and communication for biotechnology

Given biotechnology's promise for improving human well being in WCA, it is vitally important that accurate and up-to-date information be widely available to all involved in its development, application and use, as well as to the general public. Current debates on the subject reveal serious information gaps and misinformation.

The dissemination and exchange of biotechnology information is needed, using all available audiovisual and other modern information and communication technology tools and systems. Target audiences include: scientists, farmers, teachers, agricultural extension agents and farmer facilitators, policy makers, food processors and distributors, pharmaceutical producers, medical and paramedical personnel, cosmetologists, environmentalists, legal practitioners and intellectuals, development partners, journalists and other media personnel.

While modern biotechnology or the 'gene revolution' may not be a panacea for the WCA subregion's food security and economic problems, many regional bodies now consider it an important tool demanding priority attention.

Notes

1   Organisation Africaine de la Propriété Intellectuelle

References

Alhassan, W. S. (2003) 'Agrobiotechnology application in West and Central Africa', IITA, Ibadan, www.iita.org

Byerlee, D. and Fischer, K. (2002) 'Accessing modern science: Policy and institutional options for agricultural biotechnology in developing countries', *USA World Development*, vol 30 (6), pp931–948

CORAF/WECARD (2000) 'Strategic plan for agricultural research and development cooperation for West and Central Africa, 2000', CORAF/WECARD, www.coraf.org

CORAF/WECARD (2004) 'Biotechnology and biosafety project proposal', CORAF/WECARD, www.coraf.org

Maredia, M. and Howard, J. (1998) 'Facilitating seed sector transformation in Africa: Key findings from the literature', *FSII Polict synthesis No 3*, USAID, Washington and Michigan State University, East Lansing

Part Two

GM Crops for Development: the Experience of Argentina, Brazil, China, India, South Africa

6

Argentina: Adopting RR Soy, Economic Liberalization, Global Markets and Socio-economic Consequences

Daniel Chudnovsky[1]

Background

The introduction of GM crop varieties in Argentina took place in the context of structural reforms introduced in the early 1990s after a serious economic crisis. These reforms included the liberalization of trade and capital accounts and a massive privatization programme, together with a currency board. Carried out with ample international financial support, they played a major role in bringing inflation under control and regenerating economic growth. Despite growing inequality in income distribution and increasing unemployment, during the 1990s Argentina was considered the 'poster child' for the implementation of the so-called Washington Consensus.

With the external shocks that followed the Russian and Brazilian crises, the economy entered into a long recession in 1999–2000, and in 2001, into a financial, political and institutional crisis. Argentina's GDP fell by more than 20 per cent between the second quarter of 1998 and the first quarter of 2002.

The huge external debt was in default at the end of 2001 and at the beginning of 2002 the currency board was abandoned. The fall in GDP and the huge devaluation of the peso sharply reduced imports and slightly increased exports, leading to a huge surplus in the current account that mitigated the large outflows of capital during 2001 and 2002. After a very difficult 2002, in which the economy collapsed, and unemployment and poverty reached record levels, the economy began to grow again in a context of favourable international prices for commodities.

Intensification of agricultural production in Argentina during the 1990s and 2000s was one of the positive impacts of the structural reforms and of the economic policies implemented at the beginning of the 1990s. The elimination of taxes and withholdings on agricultural exports, the substantial reduction of import tariffs on inputs and capital goods, and the deregulation of some markets all created favourable macroeconomic conditions. This paved the way for a large expansion of production volumes for grains and oilseeds (from 26 million tonnes in 1988–1989 to 67 million in 2001–2002) (see Figure 6.1), and particularly for soy, which soon became Argentina's leading export item. Soy exports, which comprise not only soybeans but also soy flour and oil (see Figure 6.2), accounted for between one fifth and one fourth of total exports, depending on the evolution of international prices. The steady increase in export value occurred within a context of erratic international prices and in the face of competition from other countries, which, unlike Argentina, profit from government subsidies for production and exports.

The huge peso devaluation at the beginning of 2002 favoured the export-oriented agricultural sector. More than 70 million tonnes (half of them soy) were produced in 2002–2003, though production was somewhat reduced in 2003–2004 (see Figure 6.1).[2]

Argentina is the world's third largest producer of soy (after the US and Brazil) and a leader in the international soy export market. In 2002 it accounted for 30 per cent of that market, followed by the US and Brazil, with 25 per cent each (translating soy oil and flour into soybeans equivalents) (Ablin and Paz, 2003). The vast majority of the harvest goes to the animal feed market. The

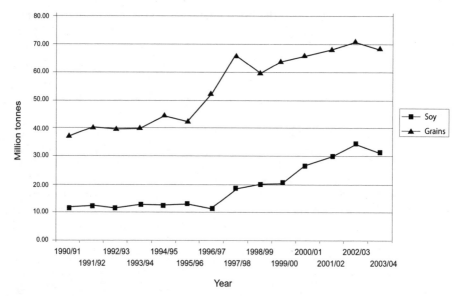

Figure 6.1 *Economic liberalization and agricultural expansion in the 1990s: Grains and soy production*
Source: Secretariat for Agriculture, Livestock, Fisheries and Food

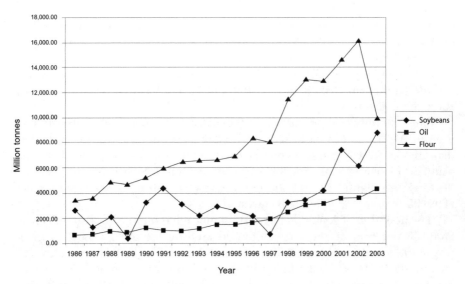

Figure 6.2 *Economic liberalization and rapid expansion of soy exports*
Source: Data from Secretariat for Agriculture, Livestock, Fisheries and Food, and CIAR

main destinations for soybeans produced in Argentina are China, the EU and Thailand, and the major buyers of soybean meal are the EU, Egypt, Malaysia and Thailand.

The government reintroduced export taxes (20–23.5 per cent) in 2002 and has maintained them until now. These taxes are a key source of financing for the national welfare plan, which the government established in 2002 to mitigate the living conditions of a substantially increased number of people living under the poverty line.

The diffusion of GM crops

Growth in cereals and oilseed production is mainly a result of the substantial expansion of the planted area (basically at the expense of livestock) brought about by a significant adoption of new technologies, notably the introduction of biotech crops. Adopting these technologies involved the procurement of machinery, equipment, fertilizers and agrochemicals (herbicides and pesticides), as well as a momentous change in terms of genetic inputs: the introduction of GM crops in Argentine agriculture.

The first GM crop commercially released into the Argentine market, in 1996, was soybean tolerant to glyphosate herbicide. Later on, GM varieties of maize and cotton, tolerant to herbicides and resistant to insects, were approved by the local authorities. The policy followed by the Argentinian government was to approve only those events that had already been approved by the EU.

When the EU authorized the import and sale of GM sweet maize (lifting its five-year-old import ban) in May 2004, Argentina approved RR maize in July of that year. And in mid-2005, it approved a new variety of maize resistant to lepidoptera and tolerant to ammonium glyphosate. This last approval took into account the technical opinion of the European Food Safety Authority, which considered that this type of maize does not have a harmful effect on human and animal health.

Since its release date, the rate of expansion of glyphosate-tolerant soybean in Argentina has increased considerably, exhibiting growth even higher than that in the US, the first country to introduce this kind of crop. The area planted to herbicide-tolerant soybean shot up from less than one per cent of the total area planted to soybeans in the 1996/97 season, to 98 per cent in the 2004/05 season (14.5 million hectares).

The adoption of lepidoptera-resistant maize has also been of significance. From 20 per cent of the total cultivated area during the 2000/01 farming season (the third year since its introduction), it reached half of the cultivated area in 2004. However, the biotech-cultivated area of maize is only 1.7 million hectares, according to James (2004). The diffusion of *Bt* cotton has, in turn, been very limited, amounting to 7–8 per cent of the total planted area in 2000–2001. Although *Bt* cotton has reached 20 per cent of the cultivated area in Argentina, only 25,000 hectares are cultivated with this type of cotton (James, 2004).

At present, Argentina ranks second, only behind the US, in terms of agricultural surface cultivated with biotech crops. It is therefore a major player in the international arena.

The sharp increase in agricultural production during the last decade has taken place hand in hand with the outstanding increase of no tillage[3] as the main farming practice for Pampas crops (favoured by the reduction in import tariffs for agricultural machinery).

This technology constituted an important factor in the expansion of production, as it introduced the cultivation of late-planted soybean (planted after the wheat harvest) to new areas. During the 1999/2000 season, for example, this translated into a virtual increment of three million hectares of arable land.[4]

Combining no till planting techniques with herbicide-tolerant soybean joins two technological concepts: new mechanical technologies that modify crop interaction with the soil, and utilization of general-use, full range herbicides (with glyphosate predominating) that are environmentally neutral due to their high effectiveness in controlling all kinds of weeds, as well as their lack of residual effect.[5]

While both factors imply a more intense use of inputs, this intensification was deemed 'virtuous' because it simultaneously lowered the consumption of herbicides with the highest toxicity level.[6] It is worth noting that throughout the period, even after the increase in the use of agrochemicals, total use per hectare of arable land was still far below that recorded in competing countries. Furthermore, the utilization of agrochemicals appears to have stabilized after the 1996/97 season. If we also consider the favourable externalities generated through the progressive recovery of soil fertility, along with other potential advantages reaped from this type of practice, such as benefits on the greenhouse effect, it seems that the overall environmental impact of these transformations was positive, at least until 2000.

Nonetheless, experts in both the National Institute of Agricultural Technology (INTA), in 2003, and the Secretariat for Agriculture, Livestock, Fisheries and Food (SAGPyA), among other analysts, have expressed serious concerns about the consequences of the soybean boom, since it has often resulted in the abandoning of crop rotation. In the Santa Fe and Cordoba provinces, where soy accounts for 80–90 per cent of the planted area, there are already serious problems of soil degradation and erosion. In the north and southwest of Buenos Aires province, soy has been expanded at the expense of maize and sunflower. Beyond the Pampas, the process of 'agriculturization' through soybean expansion in the northeast and northwest of the country is unsustainable, since these are ecologically fragile areas. Thus, an increase in land devoted to soybean cultivation could affect both the quantity and the quality of the country's natural resource endowment and might lead to a fall in agricultural production in the future.

Finally, although it is clear that the remarkable technology modernization process that took place in Argentine agriculture during the 1990s allowed a substantial increase in production and exports, it is also true that the spillovers of this process for the rest of the economy were constrained by two factors. First, the local agricultural machinery industry went through a restructuring process that implied not only plant closedowns but a strong reduction in the domestic content of locally produced machines, such as those required for no tillage situations. Second, the new chemical and genetic technological packages that are increasingly key for agricultural production are provided by a handful of multinational corporation affiliates (see below) that seldom engage in biotechnology R&D activities in Argentina. This means that the locus of technological innovation, which in previous decades was mainly in the Argentinian Pampas and carried out by public institutions such as INTA, is now carried out abroad in the headquarters of multinationals (Bisang, 2003).

The seed industry

The Argentinian seed industry dates back to the 1950s and the country also has a long-standing tradition in the field of germplasm improvement. Originally, the industry was organized around activities carried out by INTA in the public sector, a group of small local firms (such as Buck, Klein and Morgan) specializing in wheat and maize, and the subsidiaries of multinationals such as Cargill, Asgrow, Dekalb, Ciba Geigy, Monsanto, Novartis, Pioneer and Nidera.

In the late 1990s the local seed market was one of the biggest in the world and the second largest in Latin America. It reached a volume of over 1.9 million tonnes for a total value of over US$850 million in 1997, with soybean being by far the most important component, followed by maize as a distant second.

In the case of GM seeds, subsidiaries of transnational corporations generally own the genes and the technologies required for incorporating them into varieties, while local companies develop the varieties through different licensing and partnership agreements. Both types of companies are active in the conventional seed market and the subsidiaries of foreign firms are also engaged in agrochemical and other businesses (Bisang et al, forthcoming).

Biosafety and other regulations

In addition to having a significant seed industry, Argentina has enjoyed conditions favourable for the rapid adoption of GM crops. Key institutional decisions were made, particularly with regard to biosafety regulations, with the creation in 1991 of the National Advisory Committee on Agricultural Biotechnology (CONABIA) by a resolution of what was then the Secretariat of Agriculture, Livestock and Fisheries.

CONABIA is a multidisciplinary organization with advisory duties, made up of representatives from the public sector, universities and organizations in the private sector related to agricultural biotechnology. Its members perform their duties as individuals and not as representatives of the sector from which they come. The Commitee handles applications for laboratory, greenhouse testing and field trials of genetically modified plants. It also advises on issues related to trials and/or the release into the environment of GM crops and other products that may be derived from or contain them. CONABIA does not have a structure of its own. It acts through the institutions and regulations that make up the regulatory system of the agriculture sector. These are the National Institute of Seeds (INASE), created in 1991, in charge of regulating the seed trade and ensuring the availability of quality seed by preventing fraud and forgery (for which it has been vested with police powers), and the National Agricultural Food Health and Quality Service (SENASA) regulating, among other things, food safety and quality, including GM crop-derived food.

The 1973 Seeds and Phytogenetic Creations law that awards breeders rights in the creation and discovery of plant varieties adopts the principles of the International Union for the Protection of New Varieties of Plants (UPOV) 1978 Act, to which Argentina adhered formally in 1995. In this respect, there are exemptions in favour of parties that carry out plant improvements and for farmers who save seed for their own use. The legal framework was completed with the creation, in 1991, of the Argentine Association for the Protection of Plant Breeding (ARPOV), which includes all parties involved in the development of varieties. ARPOV is responsible for managing licensing agreements of varieties under the condition that seed production be conducted in Argentina.

It is worth noting that the structure of INASE was changed at the end of 2000 when it became a branch of the SAGPyA. This downgrading of the institutional status of INASE not only reduced its efficiency and flexibility but also facilitated the growth of the illegal trade in seeds (see below). In November 2003, INASE was established again as a decentralized organization with a new board of directors.

A new patent law was approved in 1995 and enacted in the year 2000. In contrast to the previous law, the new one allows the patenting of pharmaceutical products. It is consistent with the provisions of the TRIPs agreement and authorizes patenting biotechnological products and processes as long as the required patenting conditions are met (i.e. products and processes must be new, involve an inventive step and have industrial applications).

The aforementioned elements, along with the fact that Argentina constitutes a major area (amounting to 26 million hectares of cultivable land) for the potential use of new technologies outside their country of origin, provided the proper incentives and a most suitable 'landing field' for the rapid adoption of these biotechnological inputs.

Institutional factors in the diffusion of GM crops

Besides the features of the regulatory framework already mentioned, a number of idiosyncratic factors have played an important role in the diffusion of biotech crops and especially of RR soybean.

The first factor concerns the manner in which the *RR* gene was initially transferred to Argentina. Originally, access to this gene was achieved through negotiations between Asgrow and Monsanto in the US, whereby Asgrow Argentina was granted the use of the gene in its registered varieties. Later on, when Nidera acquired Asgrow Argentina, it gained access to the gene and widely disseminated it in the country. Consequently, when Monsanto sought to revalidate the patent of the gene in Argentina in 1995, it was unable to do so because it had already been 'released'.[7] However, through private settlements that expressly identify the ownership of this patent and stipulate the royalties to be paid, Monsanto was able to license the *RR* gene to other companies that have commercialized it in Argentina.[8] Therefore, conditions were never met entitling the breeder company (i.e. Monsanto) to charge the technology fee, or to restrict use of the seed by farmers, as is the case in the US.

The second factor relates to the operational aspects of the seed market and the consequent effect on the price of RR soybean. On the one hand, as mentioned, and under the UPOV Convention of 1978, farmers can legitimately keep seeds for their own use. On the other hand, there are clandestine operations (the so-called 'white bag') through which seed multipliers offer seed without the authorization of the companies holding the corresponding legal production rights. Both factors have driven down the price of RR soybean, thus promoting the rapid adoption of the technology.

Within this context, the growth of the seed market should come as no surprise, irrespective of the sharp increase observed in the surface planted with soybean, the leading crop in this market. The plateau experienced by the seed market from the years 1996/97 may be explained by the introduction of GM seed and the resulting need to obtain original seed on the part of farmers (and even clandestine seed producers). The use of white bag seed, as well as farmers' own seed, would account for the evolution of the market in subsequent years and surely had an impact on the substantial reduction in the price of RR soybean seed compared to that of conventional seed.

It should be noted that this situation is also linked to the fact that soybean seed falls into the category of autogamous species, in which genetic quality can be maintained through seed retained by farmers for their own use or used for clandestine multiplication practices. The relevance of the widespread adoption of the wheat-soybean double-cropping system during the period under analysis must likewise be taken into account as it undoubtedly constituted an additional inducement to keeping seeds for the next season.

The third factor contributing to the wide diffusion of RR soybean in Argentina was the reduction in the price of glyphosate, which stemmed from fiercer competition in local markets by dint of the introduction of new agents in the manufacturing and commercialization of this product.[9]

Compared to RR soybean, *Bt* maize and *Bt* cotton show much less dynamic performance. First, *Bt* varieties have been released much more recently, and second, farmers tend to consider these crops as a sort of insurance, yielding higher or lower profits depending on pest behaviour during each season. In addition, a technology fee, charged to farmers, is applicable to GM maize and cotton varieties. In some cases, this fee is higher than in the US. This is related to the fact that, in both crops, there are patent applications for the events involved and that, in the case of maize, it is a hybrid variety. As a consequence, farmers may not keep their own seed for planting. Therefore, the relative weight of certified seed in the corresponding market increases.

Regarding cotton, the real issue lies in the commercialization strategy. This is based on formal agreements between the sole supplying company and the farmers, whose right to 'own use' of seed is restricted. As a result, farmers have no choice but to pay four times the price of conventional varieties for *Bt* cotton seed and this hinders the diffusion of this technology in Argentina. It is clear from this that one of the main problems in Argentinian agriculture is the illegal trade of seeds, amounting to 35–50 per cent of the market.[10] Inherent in this situation are the risks of potential reductions in productivity (through the use of seed with lower genetic quality and germinative power) and risks of phyto-sanitary issues. The existence and growth of illegal practices might also mean not finding an effective way to incorporate many of the breakthroughs in biotechnology and in other conventional technologies into production. In other words, the dissemination of new knowledge is taking much longer than would be the case if the seed market worked under normal conditions.

To begin dealing with this serious situation, in July 2003 the SAGPyA passed a resolution obliging farmers to provide information on the amounts of seed used, or to be used in the next season, to plant wheat, soy and cotton. This measure appears to have contributed to an increase in the use of registered seeds, from only 20 per cent of the planted surface in 2003 to 32 per cent in 2004 (*La Nación*, 10 July 2004).

At the same time, the SAGPyA prepared a new law for discussion in parliament, proposing the establishment of a royalty to seed producers, which received strong opposition from all parties. Later on, INASE prepared legislation to form a 'fiduciary fund for technological compensation and incentive for the production of seeds'. This initiative was expected to result in a fee of 0.35–0.95 per cent of the price of the first sale of the crop being charged, initially only for soy and wheat. With the funds to be collected, a royalty would be paid to the breeders and resources would be available to finance new varieties. Furthermore, it was felt that if the farmers could prove that they were using legal seeds they might receive some compensation in the form of a reduction in taxes owed to the government.

Given the opposition to this initiative, new legislation was prepared restricting farmers' use of their own seeds to 65 hectares. Several criteria have been suggested for the paid use of seeds, with time limits according to crop, but no consensus has yet been reached among the different parties. With no agreement on the seed issue in sight, in September 2004 Monsanto decided not to charge any royalty for RR soy to licensed seed companies in Argentina for the 2004/05 season. At the same time, the company threatened to charge a US$5 per metric tonne fee on Argentinian soybean exports in the countries where this technology is patented. If exporters declined to pay the fee, they would face the prospect of being sued in the courts of European countries importing Argentinian soybeans in 2005. The Argentinian government considered this an unacceptable threat without any legal basis.

Meanwhile, several proposals made by Monsanto to the agriculture producers have been rejected. The last one included a royalty of US$1 per tonne during the 2004/05 and 2005/06 seasons, and a ceiling of up to 3 per cent of the domestic value of soy starting in the 2006/07 season and lasting until the expiration of the patent.[11]

At the end of June 2005, Monsanto started legal proceedings in The Netherlands and in Denmark to obtain a royalty on Argentinian soy exports that reached markets in those countries on ships. The Argentinian government strongly rejected the company's action and decided to participate, through the Ministry of Foreign Affairs, as an involved party in the juridical trials. Conversations among the private and public interested parties continue but their considerable differences have so far prevented an agreement on this key issue.

Research and development efforts

Public (and private) resources allocated to R&D in Argentina in all fields, including agriculture – especially biotechnology – are scarce compared to corresponding efforts at the international level. According to official data from the Secretariat of Science, Technology and Productive Innovation (SECYT), expenditure on scientific and technological activities as a percentage of GDP, in Argentina increased from 0.33 per cent in the early 1990s to 0.52 per cent in 1999. During the recession and crisis years it decreased, to only 0.44 per cent in 2002, and then rose to 0.46 per cent in 2003. R&D expenditures, after reaching 0.45 per cent of GDP in 1999, registered a decreasing trend and were only 0.39 per cent in 2002. In 2003 they increased again to reach 0.41 per cent of GDP. Agriculture accounted for 18 per cent of R&D expenditures and for a similar percentage as a field of application of R&D projects in 2003.

While financial resources are scarce, it is important to bear in mind that Argentina has sizeable human capital: 27,367 full-time researchers in 2003, most of whom were working in public universities and institutes. Regarding biotechnology, according to a 2000 survey of 18 research organizations (out of 41 contacted) by the International Service for National Agricultural Research, the country had about 12 organizations with major capabilities in molecular biology and generic engineering that employed approximately 300 researchers. Resources devoted to research were US$3.5 million, excluding researchers' salaries. These organizations were mainly in the public sector and in public universities. Major laboratories include the INTA's Castelar Institute (Biotechnology, Genetics, Plant Physiology, Veterinary Sciences); those located at the Institutes of Genetics and Biotechnology (INGEBI-Conicet), the Institute of Biochemical Research at the Campomar Foundation, the Centre of Photosynthetic and Biochemical Studies (CEFOBI-Conicet) and the Centre of Animal Virology (CEVAN-Conicet); those of some public universities such as Universidad Nacional de la Plata, Universidad Nacional de San Martin and Universidad Nacional de Tucumán; and the agricultural division of Bio Sidus, in the private sector (Cohen et al, 2001).

An ongoing survey of biotechnology firms in Argentina indicates that in 2002–2003, the 71 firms surveyed devoted 0.9 per cent of their sales revenues to R&D. In agricultural biotech, seed firms assigned 0.52 per cent of their sales revenues to R&D (Bisang et al, forthcoming). The low figures that both public and private institutions devote to R&D in biotechnology are in dramatic contrast to the resources assigned to this field not only in the US but also in developing countries such as China and Brazil.

There is also great diversity in the focus of research. According to an ISNAR survey, agriculture-related applications include the diagnosis of phytopathogens in several crops, the development of biological control agents, and the use of micro-propagation techniques, molecular markers and genetic engineering of different crops such as garlic, onion, potatoes, sunflower, maize, wheat, alfalfa, strawberry, tomato, rye, citrus, cranberry, sugar cane and yerba mate (Cohen et al, 2001).

Table 6.1 *Permits for the release of GM crops into the environment (field trials) by type of organization*

	1991/93	1994	1995	1996	1997	1998	1999	2000	2001	2002	2003	2004	Total
Transnational corporations	11	17	26	28	62	65	70	52	49	62	93	96	631
Local companies	8	4	6	6	12	12	10	10	4	4	5	19	100
Government agencies	2		4	6	4	13	1		8	1	1	5	45
Universities								3	2	3		1	9
Total	21	21	36	40	78	90	81	65	63	70	99	121	785

Source: Own elaboration on data of CONABIA

Table 6.2 *Permits for the release of GM crops into the environment (field trials) by type of crop*

	1991/93	1994	1995	1996	1997	1998	1999	2000	2001	2002	2003	2004	Total
Soybean	3	5	9	5	7	12	10	15	10	11	18	13	118
Maize	8	10	18	23	41	40	44	22	23	42	58	96	425
Cotton	4	2	5	4	7	4	5	9	8	3	1	2	54
Wheat	1		1	2	2	2	1	3	3	1		3	19
Sunflower		2		2	17	24	18	7	4	3	1		78
Potato			1	1	2	3	1	4	3			5	20
Alfalfa					1	4		1	8	1			15
Others	5	2	2	3	1	1	2	5	4	9	22	4	60
Total	21	21	36	40	78	90	81	66	63	70	100	123	789

Source: Own elaboration from data of CONABIA

In the recently approved Strategic Plan 2005–2015 for the development of agricultural biotechnology (Ministerio de Economía y Producción, 2004), it is pointed out that although the country has good research capacity in life sciences, as reflected in the three Nobel Prizes obtained by Argentinian researchers, this is not the case in modern biotechnology. With relatively few biotech plants and related industrial developments in the country, the demand for professionals in this area has been limited. Training of additional human resources in biotechnology is required. However, it should be noted that the Argentinian biotech industry has important capabilities regarding access to information, lab techniques, modern equipment and participation in international networks. It employs agricultural biotech tools and has excellent facilities for improving and adapting new plant varieties. Nonetheless, beyond their meaningful contribution to R&D activities on some crops (such as alfalfa and potato) and in the sphere of veterinary science, institutes devoted to agricultural biotechnology research in Argentina have hardly participated in the events approved by CONABIA.

As shown in Table 6.1, in Argentina, as well as in many other countries, multinational corporations have taken the lead in the process of environmental field trials. For the most part these field trials have focused on maize and soybean (see Table 6.2). Given that Argentinian crop growing areas are analogous to those in the northern hemisphere, for which the technologies were originally developed, the only costs that multinationals have to bear are those of backcrossing the new genes into well-adapted already existing varieties – a process that is much simpler than the actual development of a GM crop variety.

As already mentioned there are a number of local private breeders that have been able to maintain their businesses through partnerships with multinational corporation affiliates; the latter provide the GM genes, which are combined with varieties well-adapted to local conditions, owned by local breeders (Trigo et al, 2002).

Economic and social impacts

There is good information on the distribution of benefits of biotech crops between input suppliers and farmers based on our own research (Trigo et al, 2002).[12] Some data are available concerning the growing concentration on agricultural production and on employment creation, but research is required to analyse how this relates to biotech crop diffusion.

Distribution of benefits among farmers and input suppliers

In the case of RR soybean adoption, benefits are derived from reductions in production costs and from the impact these reductions have had over the areas planted with soybeans. The extra income (which, in the absence of this technology, would not have been generated until the year 2001–2002) was estimated as US\$5.2 billion (by comparing two alternative scenarios with and without RR soy). Following the same methodology, in the case of *Bt* maize the benefits were estimated as US\$400 million, and in *Bt* cotton as US\$40 million.

Most of the benefits of RR soybean adoption ended up in the hands of farmers, who captured more than 80 per cent of the total, mostly due to an increase in production. The benefits to input suppliers have been relatively low. If the white bag had existed, input suppliers would have increased their share in the overall benefits from 13 per cent to 18 per cent (see Table 6.3). Nevertheless, the lion's share of the benefits would have accrued to farmers rather than to input suppliers.

The evidence available for *Bt* maize and *Bt* cotton (see Table 6.3) does not point in the same direction. Most of the benefits have accrued to input suppliers in the case of both *Bt* crops. With *Bt* maize, where protection comes from the hybrid nature of the seed, the importance of IPRs is clearly shown. In *Bt* cotton, Quaim (2002) shows that adoption rates would have been much higher if the technology supplier's seed pricing policies had been more flexible.

Table 6.3 *GM crops in Argentinian agriculture: Distribution of gross accumulated benefits (%)*

	Case	Farmers	Input providers
Soybean RR	Without 'white bag'	82	18
	With 'white bag'	87	13
Cotton *Bt*		17	83
Maize *Bt*		21	79

Source: Trigo et al (2002)

Increasing concentration of production

Available information indicates that there has been a strong trend towards land concentration. A study carried out by Mora y Araujo (2000) showed that between 1992 and 1999, the number of farms dropped from 170,000 to 116,000 (a 32 per cent reduction). At the same time, there had been an increase in the median size of farms, from 243 to 357 hectares.

The classification by size (small, medium and large) proposed by Pucciarelli (1997) takes into account a different set of parameters and is more precise than the grouping of farms by area. Analysis according to this classification indicates that, from 1993 to 1999, small farms decreased from 85 to 69 per cent; medium ones increased from 9 to 18 per cent; and large ones, from 6 to 13 per cent. In the same manner, by 1999, the area accounted for by each group was: small farms, 28 per cent; medium farms, 23 per cent; and large farms, 49 per cent.[13]

Data from the Agricultural Census of 2002 complete the picture that emerged from previous studies. Comparing 2002 data with that of the previous census in 1998, in the country as a whole, the number of farms decreased from 421,000 to 332,000 (a 21 per cent fall) and the average surface of the unit increased from 421 hectares to 518 hectares. In the Pampas, the reduction in the number of farms was sharper (29 per cent) and the increase in the average surface higher (36 per cent) than in the rest of the country (Lazzarini, 2004).

According to some experts, the reduction in the number of farms – mainly of small and medium sizes – is associated with the phenomenal expansion of RR soy. Furthermore, the surviving farms are forced to buy their seeds and agrochemicals from the affiliates of multinational corporations (Teubal, 2003). Unfortunately, this association and its causes have not been properly tested and no references are made by Teubal either to reductions in the prices of seeds and agrochemicals or to the illegal trade in seeds.

It has also been argued that the incorporation of new technologies led to huge levels of indebtedness; hence many small farmers were not able to continue with their production activities because of the debt burden (Bisang, 2003). While it is true that the machinery required for no till farming is expensive and hence only farmers with large farms can afford to buy it, most farmers with medium- and small-size farms can rent the machinery from specialized firms that provide such equipment. Furthermore, in a microeconomic study on RR soy diffusion in Argentina, Qaim and Traxler (forthcoming) did not find any adverse effect of the new technology on small-scale farmers.

In the case of *Bt* cotton, we found that farmers with low technological capacity (generally small-scale ones) received 12 per cent of the benefits and accounted for 27 per cent of the planted surface. Regarding *Bt* maize, farmers with low technological capacity received 13 per cent of the benefits and cultivated 21 per cent of the land. Although the more expensive seeds are a clear disadvantage for small-scale farmers, access to them could be facilitated through trade credit, though this type of credit is generally expensive.

In contrast, access to specialized agricultural equipment would require long-term credit lines and a change in the production process. As mentioned above, the development of firms specialized in renting machinery to small- and medium-scale farmers may have mitigated this problem. In any case, further research on the impact of the diffusion of biotech crops on small- and medium-scale farmers is required.

Employment trends

The diffusion of labour-saving technologies through increased mechanization and use of tractors, plus (over the last three decades) an increase in the average power of equipment, while facilitating economies of scale, has had an impact on rural employment. The number of direct jobs in the sector fell from 1.86 million in 1926 to 783,000 in 1993.[14] However, since 1993 the trend has significantly reverted, to reach 966,000 jobs in 1999 (latest available year).[15] This positive difference of nearly 200,000 jobs is likely to have been the result of the simultaneous processes of agriculturization (crops substituting for livestock) and intensification of production systems based on biotech crops.[16]

The introduction and rapid expansion of late-planted soybean (planted after the wheat harvest) has played a substantial role. In the 1999/2000 season, this practice implied a virtual increase of 3 million hectares of arable land, and thus, an additional demand for labour. The most remarkable aspect is that this increase in the level of employment took place simultaneously with an increase in the partial productivity of labour in the primary sector of 3.26 per cent per year for the period 1990–1997, *and* an almost five point increase in total unemployment in Argentina. Thus, the technological package seems to have had positive effects from a social perspective, at least in terms of job creation (Trigo and Cap, 2003).

Concluding remarks

Keeping in mind that so far Argentina has encountered no difficulties in accessing target markets for its RR soybean exports and that, in spite of the perceptions of foreign consumers, price differentials between conventional and RR soybeans on the world market do not penalize them, it is hardly surprising that almost the whole Argentine soybean crop is RR.

Neither is it surprising that not only input suppliers but also farmers, the scientific community and government authorities are all in favour of this new technology, as clearly reflected in the collective preparation of the Strategic Plan 2005–2015 led by the recently created Office of Biotechnology.

According to the first national survey on public perception of science, carried out in 2004 by the Secretary of Science and Technology, most of the population accepts the agricultural use of biotechnology. The main concern is access to healthy and sufficient food. Only a few NGOs, especially Greenpeace, have introduced themselves as part of the international debate in Argentina. It is also true that in some provinces, due partly to local NGOs, legislation is under discussion to make labelling of GM crops compulsory.

Nonetheless, in taking a long-term view, the extraordinary success of RR soybean in Argentina should be regarded with more caution. First, pushed by high international prices, excessive reliance on this crop may affect the fertility of the soil. In this connection, a 2003 report by INTA states that the no tillage system plus RR soybean cannot continue as a sustainable strategy without rotating crops in the Pampas. At the same time, the agriculturization process caused by soybean expansion is not sustainable in the ecologically fragile northeast and west areas of the country. Both processes could affect the quantity as well as the quality of the country's natural resources and lead to a fall in agricultural production.

Although according to INTA (2003) more sustainable production methods (based on rotation with maize and livestock) are available and are being adopted by some farmers despite their higher operating costs, the fact that 50 per cent of the land is leased and the price of the lease is fixed in kilograms of soybeans is a serious constraint for the diffusion of these methods. While the recent release of new GM maize varieties and lower international prices for soybean may make it more attractive for farmers to plant less soy and more maize, the mono-cropping issue is a very difficult one. Beyond many discussions in technical meetings, no serious attempt to deal with it at a political level is visible.

Second, the trend towards a differentiated world market for GM and non-GM products in view of increasing eco-labelling requirements in import markets (to meet consumer fears about these types of food items) may negatively affect the prices at which Argentinian producers sell their oil seeds. The way the WTO finally proceeds in the conflict regarding the GM crop moratorium between the European Union on the one hand, and the United States, Canada, Argentina and other GM product producing countries on the other hand, may also influence Argentinian exports to the world market.

Third, the difficulties in finding a compromise solution with Monsanto on the fee for the RR technology, and generally on finding incentives for the seed industry development and measures to curb the illegal trade in seeds, are clear indications of the institutional problems Argentina is facing with regard to making further advances in this technology. Furthermore, the relatively low hierarchy of the legal and institutional framework of CONABIA and the limited research capacity on biosafety are important issues to be addressed (Burachick and Traynor, 2001).

At the same time, the domestic policy process regarding biotechnology should be re-examined. It will be important to analyse ways in which leading stakeholders can become more responsible for the long-term effects of technology advances and to enable more participation on their part. For example, parliament, consumers and environmental NGOs have been largely ignored in the policy process. In this connection, an important initiative mentioned in the Strategic Plan for Biotechnology is to establish public hearings before new events are approved for commercialization.

Fourth, possibilities should be considered of increasing financial and human resources so as to allow a greater participation of local firms and institutions in the research and monitoring processes, which have so far been influenced by affiliates of foreign companies. While this is precisely one of the main points of the Strategic Plan for Biotechnology, it is not clear how the good intentions of this plan are going to be translated into actual financial resources for research, and for increasing the training and development of human resources in this critical area. It is also surprising that neither research priorities for local efforts nor specific ways to obtain positive spillovers from the activities of foreign firms in Argentina are discussed or even mentioned in the Strategic Plan.

Fifth, more research is required on the socio-economic impact of biotech crops. The impact of this technology on the growing concentration of production and especially on small- and medium-scale farmers is a key issue on which little research is available. The distribution of benefits between input suppliers and farmers should also be further studied in soy, maize and cotton, especially in view of the changes in relative prices that have taken place since the peso devaluation in 2002 and the existance of withholding taxes.

Finally, more research is also required on the environmental impact of biotech crops in Argentina, taking into account not only the above-mentioned mono-cropping issue but also the excessive reliance on glyphosate, the growing use of fertilizers and the impacts on health and on biodiversity.

Notes

1   Comments from Sakiko Fukuda-Parr and other participants at the Bellagio Conference are gratefully acknowledged. The usual caveat applies.
2   According to the last official estimate, a historical record of 84 million tonnes will be produced in 2004/05 of which soy will account for 38.3 million tonnes.
3   No tillage maintains a permanent or semi-permanent organic soil cover (e.g. a growing crop or dead mulch) that protects the soil from sun, rain and wind and allows soil micro-organisms and fauna to take on the task of 'tilling' and soil nutrient balancing – natural processes disturbed by mechanical tillage.
4   According to Benbrook (2005), of the 5.6 million hectares of land newly planted to soybeans in 1996–2003, 25 per cent has come from conversion of cropland growing wheat, maize, sunflower and sorghum, 41 per cent from conversion of forests, and 27 per cent from conversion of former pastures.
5   According to the classification of pesticides by hazard prepared by the WHO, glyphosate falls into the category of herbicides of toxicity class IV, which are the most benign ones. Qaim and Traxler (forthcoming) show that the adoption of RR soybeans in Argentina has led to an 83 per cent reduction in the use of herbicides with toxicity class II, and that herbicides with toxicity class III have been phased out.
6   However, Benbrook (2005) points out that reliance year after year on a single herbicide accelerates the emergence of genetically resistant weed phenotypes and this is, according to him, what is going to happen in Argentina.
7   Monsanto states that the patent request made in 1995 was rejected in 2001 by the Supreme Court of Justice in a polemical interpretation of the validity of the revalidated patent request and the entry into force of the new patent law. Monsanto considered that this interpretation not only affected it, but also dozens of companies that requested revalidated patents (www. monsanto.com.ar/tecnologiarr). According to the Argentinian government, Monsanto made the request to revalidate the patent after the 12 months

allowed by the law in force.

8   Monsanto has started to commercialize its own RR varieties since 1999.

9   Monsanto also made the glyphosate in its Argentinian manufacturing plant and filed an antidumping procedure against imports of that product from China. After a year and half, in February 2004 the government decided against Monsanto's claim and its decision was widely supported by the agriculture sector.

10   The illegal seeds market (50 per cent according to the company, in contrast to 18 per cent of the seeds certified and 32 per cent self-produced by the farmers) was the main reason that led Monsanto to close its soybean business in Argentina at the end of 2003, arguing that it was not profitable (*La Nación*, 21 December 2003). It is likely that the news of an adverse decision on glyphosate imports also influenced the company.

11   See www.monsanto.com.ar/tecnologiarr.

12   Qaim and Traxler (forthcoming) estimate the benefits with a different methodology. They include benefits to consumers, which are very low compared to those accruing to producers.

13   Mora y Araujo's work uses the 'farm' concept to refer to those production units under the same management and not necessarily the same ownership, since there has been a remarkable increase of farm renting.

14   This reduction of nearly one million jobs had a negative aspect – the laying off of labour (socially undesirable effect) – and a positive one as well – an amazing increase in the productivity of workers, made possible by modern mechanical technologies, a fact that enabled the sector to maintain its international competitiveness throughout the 20th century.

15   According to a study by Llach et al (2004) based on the Input-Output table of the Argentine economy for 1997, employment in the primary agricultural sector increased by more than 270 thousand.

16   It is possible that part of the increased labour demand is due to labour requirements in areas outside the Pampas or in other crops. Information yet to be processed from the latest Agriculture Census would shed light on this issue.

References

Ablin, E. and Paz, S. (2003) *El Mercado Mundial de Soja: La República Argentina y Los Organismos Genéticamente Modificados*, Centro de Economía Internacional, Trade and International Economic Relations Secretariat, Ministry of Foreign Affairs, International Trade and Worship, Buenos Aires

Benbrook, C. (2005) 'Rust, resistance, run down sils, and rising costs: Problems facing soybean producers in Argentina', Technical Paper No 8, January, Ag BioTech InfoNet

Bisang, R. (2003) 'Apertura económica, innovación y estructura productiva: La aplicación de biotecnología en la producción agrícola pampeana

Argentina', *Desarrollo Económico*, vol 43 (71), Buenos Aires

Bisang, R., Diaz, A. and Gutman, G. (forthcoming) 'Empresas biotecnológicas en Argentina', (mimeo)

Burachik, M. and Traynor, P. (2001) *Commercializing Agricultural Biotechnology Products in Argentina: Analysis of Biosafety Procedures*, International Service for National Agricultural Research (ISNAR), The Hague, The Netherlands

Cohen, J., Koimen, J. and Verástegui. J. (2001) 'Plant biotechnology research in Latin American Countries: Overview, strategies and development policies', prepared for IV Latin American Plant Biotechnology Meeting, REDBIO, Goiania, Brazil

INTA (2003) *El INTA Ante la Preocupación por la Sustentabilidad de la Producción Agropecuaria Argentina*, INTA, Buenos Aires

James, C. (2004) 'Global status of commercialized biotech/GM crops', ISAAA Briefs No 32, The International Service for the Acquisition of Agri-biotech Applications, Manila

Lazzarini, A. (2004) 'Avances en el análisis del CNA 2002 y su comparación con el CNA 1988', paper prepared for the project Sistematización y Análisis del Censo Nacional Agropecuario 2002, Instituto de Economía y Sociología, INTA, Buenos Aires

Llach, J., Harriague, M. and O'Connor, E. (2004) 'La Generación de empleo en las cadenas agroindustriales', Fundación Producir Conservando, Buenos Aires

Ministerio de Economía y Producción (2004) 'Plan estratégico para el desarrollo de la biotecnología agropecuaria 2005–2015', Secretaria de Agricultura, Ganadería, Pesca y Alimentos, Oficina de Biotecnología, Buenos Aires

Mora y Araujo, M. (2000) 'Perfil productivo de la pampa humeda', Bolsa de Cereales de Buenos Aires, available at www.bolsadecereales.com

Pucciarelli, A. (1997) 'Estructura agraria de la pampa bonaerense: Los tipos de explotaciones predominantes en la provincia de Buenos Aires', in Barsky, O. and Pucciarelli, A. (eds) *El Agro Pampeano: El Fin de un Período*, FLACSO, Universidad de Buenos Aires, CBA Publications Office, Buenos Aires

Qaim, M. and Traxler, G. (forthcoming) 'Roundup ready soybeans in Argentina: Farm level environmental and aggregate welfare effects', accepted for publication in *Agricultural Economics*

Teubal, M. (2003) 'Soja transgénica y crisis del modelo agroalimentario argentino', *Realidad Económica*, no 196, Buenos Aires

Trigo, E. and Cap, E. (2003) 'The impact of the introduction of transgenics crops in Argentinean agriculture', *AgBioForum*, vol 6 (3), University of Missouri, Columbia

Trigo, E., Chudnovsky, D., Cap, E. and López, A. (2002) *Los Transgénicos en la Agricultura Argentina: Una Historia con Final Abierto*, Libros del Zorzal/IICA, Buenos Aires

7

Brazil: Confronting the Challenges of Global Competition and Protecting Biodiversity

*José Maria F. J. da Silveira and
Izaías de Carvalho Borges*

Introduction

Brazil has become a major player in world agriculture over the past 25 years thanks to scientific research and technological development that have contributed to increased productivity, better quality products and more diversified production. Adoption of agricultural biotechnology has been part of this trend; with 9.4 million hectares of herbicide tolerant soybean varieties in 2005, the country ranked third among those cultivating GM crops (James, 2005). Biotechnology research follows global trends but this technology has also raised concerns about increasing concentration in the seed market, rising prices and ecosystem risks. These aspects need strong management in Brazil given the importance of biodiversity and the country's socio-economic structure. Identity preservation and legal issues related to biosafety are also central because they affect Brazil's position as a leader in biodiversity preservation, sustainable agriculture and competitive agribusiness. This chapter documents Brazil's experience in agricultural biotechnology.

The first section provides an overview of Brazil's recent biotechnology research and the second analyses the evolution of institutional structures. The third addresses trade issues and market trends affecting cultivation of GM crops and gives a brief analysis of the social and economic impacts of their introduction. The last section offers main conclusions.

Modern biotechnology research:
Activities and achievements

Brazil's scientific community has led the development of agricultural biotechnology, together with private enterprises and government agencies. While 'new', this scientific and technological development is also an extension of the plant and animal breeding of the last 20 years, as well as the country's long history of agricultural research dating back to the 19th century.

Overview

Much of Brazil's biotechnology research is conducted by public research institutions, while public universities and private companies also play key roles.[1] While cooperation between public and private sectors is increasing, research on global (platform) crops is performed by multinationals at the same time as tropical foods, including sugar cane, beans and manioc are developed by public institutions, private research institutions (sponsored by cooperatives) and small biotechnology firms. Multinational corporations such as Monsanto, Syngenta, Dow, Bayer and Dupont diffuse genes in platform crops by inserting them in their own varieties, or in varieties from local seed companies' products, backcrossing genes with local material.

Strategic programmes for development of GM crops underway in Brazil focus on soybeans, rice, potatoes and maize. Other priorities include beans, eucalyptus, papaya and sugarcane. These programmes target herbicide tolerance and resistance to plagues and diseases, adaptation to adverse environmental conditions, and nutritional and pharmaceutical properties.[2] Table 7.1 summarizes the main research on tropical, sub-temperate and platform crops. National public institutions focus on the tropical and sub-temperate crops and on adapting cultivars to meet local constraints, such as inadequate insect resistance, going beyond backcrossing genes to locally adapted varieties. The strategic policy of the Brazilian Corporation for Farming and Livestock Research (EMBRAPA) and others is to cooperate rather than compete with private corporations in developing commercial products (Lima, 2005).

Biosafety, a relatively new area of research, requires stronger support and capacity building. Brazil was one of the first Latin American countries to invest in good laboratory practices (GLP) that make possible risk analyses of GM products in public labs. The GLP network comprises more than 100 scientists from fields covering nearly all multidisciplinary aspects of GMO risk assessment. In 2002, EMBRAPA launched BIOSEG, a four-year project to develop biosafety protocols adapted to Brazilian conditions, to be applied through five case studies.[3]

Table 7.1 *GM research by Brazilian institutes (since 1995)*

	Field trials		Laboratory study	
	Trait	**Company/ institution**	**Trait**	**Company/ institution**
Tropical and sub-tropical crops				
Bean	glufosinate tolerance, and resistance to golden mosaic virus	EMBRAPA		
Banana			Resistance to fungi	EMBRAPA
Cocoa			Resistance to fungi	Unicamp, Ceplac, EMBRAPA e Uesc
Eucalyptus				Esalq/USP
Passion fruit	Disease resistance	Esalq/USP		
Papaya	Papaya – PRSV virus resistance	EMBRAPA		
Sugar cane	Herbicide resistance, Lepidoptera resistance, and resistance to SCMV (yellow) virus	EMBRAPA, IAC, Coopersucar, CTC	Resistance to borer or Lepidoptera	Ufscar, Esalq, IAC, Coopersucar
Tobacco	Resistance to TSWV and PVY virus	EMBRAPA		
Temperate crops				
Barley				State University of Maringa
Canteloup				
Carrot	Isolate carotenoid genes	EMBRAPA		

Lettuce				EMBRAPA
Potato	Potato resistance to PVY and PLRV virus	EMBRAPA		
Tomato	Resistance to Gemini and tospovirus	EMBRAPA		

Platform crops

Cotton	Glyphosate tolerance, Lepidoptera resistance, multiple resistance	Facual- Famato, EMBRAPA, Delta Pine		
Maize	Herbicide tolerance, Lepidoptera resistance, multiple resistance	EMBRAPA	Aluminium and phosphorous deficiency	EMBRAPA
Rice	Glufosinate tolerance		Salt tolerance and resistance to fungi	EMBRAPA
Soybean	Glufosinate and Imidazoline resistance, Lepidoptera resistance	EMBRAPA	Insect resistance	EMBRAPA

Source: EMBRAPA (2004); Fonseca et al (2004)

Genomic and proteomic research

In 1997 the Foundation of Research Support of the State of São Paulo (FAPESP) organized the Organization for Nucleotide Sequencing and Analysis (ONSA) Network, a virtual genomics institute initially comprising 30 laboratories linked to research institutions in the state. Its first project, completed in 1999, deciphered the genetic material of the *Xylella fastidiosa* bacteria that causes citrus variegated chlorosis, and Brazil made history as the first country to obtain sequencing of a phytopatogen – an organism that causes disease in an economically relevant plant.

Subsequently, sequencing work on *Xylella* led to two sub-projects: Functional Genome of the *Xylella*, and Agronomic and Environmental Genome, which studies agronomically relevant micro-organisms. By 2003, 14 genome projects funded by FAPESP were ongoing or completed, nine focusing on issues relevant for Brazilian agriculture.

Private companies take part in practically all of the genomic studies networks:

- The *Xylella* Genome Project, a partnership between FAPESP and Fundecitrus, a private organization representing citrus producers, involved some US$15 million, of which 3.2 per cent was contributed by the private sector (Dalpoz et al, 2004);
- The SUCEST Genome Project for sequencing sugar cane benefited from the participation of Coopersucar, an agro-industrial organization equipped with R&D laboratories, which conducted sequencing and data-mining;
- The MCT-financed Genolyptus Project, a partnership of seven universities, EMBRAPA and 12 private companies, aims to achieve productivity gains, reduce industrial pollution and increase Brazil's competitive edge in the pulp and paper market. The project's first stage involves a US$4.1 million investment, 70 per cent from public funds and 30 per cent from private initiatives (Dalpoz et al, 2004).[4]

The Brazilian Genome Project and the regional genome projects are also significant initiatives, but focus on health and industrial applications.

Brazilian 'in house' agricultural biotechnology: Some considerations

Recent agricultural research programme evaluations (Furtado et al, 2004) indicate that capacity development has been the main outcome of investment in genomics. New GM crop varieties are awaiting approval by National Biosafety Technical Committee (CTNBio) for commercialization, including new papaya varieties.[5] Breeding programmes have achieved improvements in transgenic varieties of eucalyptus and sugar cane (tolerance to herbicides and resistance to insects).

Brazil's ability to introduce new genes in its breeding programmes shows that its research can go beyond the introduction of genes by applying backcrossing techniques. Results obtained with the sugar cane genome indicate that firms with venture capital funding might discover new genes, and that the main obstacles stem from gene patents not being acknowledged.

Lima (2005) has shown that the strategic function of state-run corporations – under the leadership of EMBRAPA – is to provide scientific support to small Brazilian companies, cooperatives and research associations working on development and diffusion of new cultivars, especially of tropical crops not targeted by multinationals. These companies must keep pace with evolving agricultural technology, not only to be able to negotiate with the major seed corporations from a vantage point, but also to develop identification protocols and risk monitoring procedures, thereby reducing current GM seed diffusion transaction costs.

Institutional environment

Parallel to scientific research, development of an institutional framework for modern biotechnology has been an important strategy. Although most research has been conducted by public institutions, an increase in the participation of private organizations is evident (Salles-Filho, 2001).

Research infrastructure and financing

Public sector R&D activities are financed by federal and state agencies and resources, particularly the National Research and Development Council (CNPq)[6] and FAPESP.[7] FAPESP has created the Genome Project, the Biota Project (Biodiversity), the Structural Molecular Biology Network and the Virus Biological Diversity Network. The first three projects focus on agriculture while the last one concerns health. They receive about US$10 million per year, equivalent to 4 per cent of FAPESP's total 1997–2003 budget.

Recently, Congress approved Law No. 0,332/2001, instituted by the Biotechnology Sector Fund (CT-Biotechnology) to provide resources for R&D activities in biogenetics and genetic resources. CT-Biotechnology receives 7.5 per cent of the 'Intervention of Economic Domain Contribution', intended to strengthen competence in biotechnology through partnerships among teaching, scientific research and technological development institutions and private enterprises.[8]

The key institution involved is EMBRAPA, which coordinates the National System for Research in Farming and Cattle-Raising (SNPA) in cooperation with other research institutions and universities. Agreements with private national and multinational corporations provide access to enabling technologies and result in new products, mainly in tropical and sub-tropical crops such as sugar cane, papaya and beans.[9] Recognized worldwide, EMBRAPA is Brazil's major centre for tropical agriculture and livestock technology. EMBRAPA's budget – approximately US$300 million per year, excluding wages and infrastructure costs – has remained stable since 1994, including expenses on personnel, operations and capital. The annual budget for agricultural biotechnology is estimated at about US$30 million, not including wages and fixed costs. A major concern is how EMBRAPA's innovations might benefit large numbers of farm families. In the debate on how agricultural biotechnology can help solve problems of rural inequality and poverty, social movements, particularly the Landless Workers' Movement (MST) have declared a preference for agro-ecological techniques as a viable alternative to GM crops. However, since 2000 some MST settlements in South Brazil have adopted GM soybean varieties (Lima, 2005).

Agricultural biotechnology spurs intensive collaboration among universities, SNPA and EMBRAPA's key agricultural biotechnology departments. In fact, a major portion of Brazil's laboratories, equipment, human resources and experimental fields are in universities, which act as important links between basic research and the market; many biotechnology companies sprang from research conducted at university laboratories.[10]

Training of human resources

Modern biotechnology demands professionals not only from traditional biology, agronomy and health sciences, but also from informatics, bioprocess engineering and technological, commercial and financial management. Most of Brazil's scientific researchers are trained at publicly financed universities. Life sciences have received the most funds – a share rising from 38.7 per cent to 40.8 per cent from 1998 to 2003. Within this field, of the total funds available, biological and agrarian sciences received 18.3 per cent and 14.4 per cent, respectively (MCT/CNPq, 2004).

A number of scientific research groups have been organized throughout the country while human resources were trained to work with new technologies. In 2000, 6616 biotechnological research endeavours were spread among 1718 public groups (760 in federal and 460 in state universities), involving 3814 lines of research, of which 1075 were in agrarian sciences (Salles-Filho, 2001; Batalha et al, 2004).

Despite these efforts and many graduate courses, a recent survey indicated a lack of qualified professionals for research in important areas including bioprocess engineering,[11] genetic sequencing, legal assistance in environmental and intellectual property issues, valuation of biodiversity and administrative and financial management (Batalha et al, 2004).

Intellectual property rights and GM crops

The Brazilian intellectual property rights system set up in the 1990s 'has attempted to respond to the requirements of TRIPS and likewise to meet the demands defined by the Biological Diversity Convention' (Dalpoz et al, 2004).[12] Plant biotechnology property rights are regulated by the Law of Patents, enacted in 1996, and the Law of Cultivars Protection, enacted in 1997. These laws provide for 'the appropriation of innovations, and ensure intellectual property on cultivars, creating openings for the collection of royalties and technology fees' (Fuck, 2005). They stimulated a wave of mergers and takeovers of seed companies in the 1990s, increasing multinational corporations' participation in the Brazilian seed market (see below).

In the case of GM crops, current Brazilian law does not acknowledge patenting of genes. While a new plant created by insertion of genes can be patented, a patent can only be accepted on insertion of the gene. Thus, 'a patent covering the insertion of the gene provides guarantees to the effect that the farmer cannot reproduce transgenic seeds without the authorization of the holder of the patent' (Fuck, 2005).

Two further possibilities exist for harmonizing access to technology and the protection of property rights. First, the Law of Industrial Property provides the possibility of licensing a gene for insertion in third party plants. Second, it is also possible to execute joint technology transfer agreements. According to Fuck (2005), this 'is one of the forms of access to state-of-the-art technology which unites local R&D efforts and the transfer of know-how generated abroad'. EMBRAPA, for instance, has such a technology transfer agreement with Monsanto for GM organisms (Fuck, 2005).

The new patent regime has also stimulated partnerships between private companies and state-run companies, especially EMBRAPA. It has also created a new relationship between EMBRAPA and the seed production industry, 'which began to invest in the generation of cultivars in exchange for exclusivity in the production and marketing of the resulting seeds over a given period of time' (Fuck, 2005). For example, EMBRAPA entered into agreement with Monsanto research projects for RR soy, and for a marketing of GM soy under its exclusive brand name.

The biosafety law

Like that of the European Union, Brazil's biosafety law is based on the 'precautionary principle', in contrast to the US approach based on the 'principle of substantial equivalence'. The 1995 Biosafety Bill (Law No. 8974) was approved to regulate all activities and stages of biotechnology, from fundamental research, through experimentation and planting, up to handling, transportation, marketing, storage and dissemination. The law set up the CTNBio as a body under the Ministry of Science and Technology to handle regulation.

This arrangement, giving CTNBio authority to issue final opinions on environmental issues, created conflict with other environmental regulating bodies, including the Ministry of Environmental Affairs (MMA) and the state governments' departments of environmental affairs. Sectors of society opposing biotechnology deemed the power conferred on CTNBio to represent a possible weakening of the MMA and other bodies covering GM crops.

This conflict effectively hindered enforcement of the Biosafety Law. In 1998, CTNBio issued an opinion favouring clearance of Monsanto RR soybeans for marketing in Brazil. In response, to thwart the sale of RR soybeans, the Consumers Defence Institute (IDEC) and Greenpeace took civil action disputing CTNBio's technical opinion. The courts granted an injunction, making the clearance of GM soybeans conditional upon an environmental impact survey. According to Monteiro (2003) the court cited the Federal Constitution (Article 225, §1) requiring that public authorities demand a prior environmental impact study for any activity representing a potential risk for serious environmental damage. It also deemed that CTNBio was not the competent 'public authority' to require such a study, much less to exempt anyone from conducting one.

The production and marketing of RR soybeans in Brazil was prohibited from 1998 as a result of this injunction. But this did not stop a majority of farmers in the State of Rio Grande do Sul from illegally importing RR seeds from Argentina. The clandestine dissemination of RR soybeans reached such proportions that the Federal Administration had to enact a Provisional Measure permitting their use up to 2003/2004.

Coping with several regulatory bodies creates other obstacles for GM plant research while the highly complex bureaucratic system's tangle of laws and provisions increases research time and costs. If a company desires to obtain approval for its research it must:

1   obtain a Biosafety Quality Certificate;
2   procure authorization to conduct research from the Ministry of Agriculture, Cattle-Raising and Provisioning (MAPA) in the case of agricultural products;
3   obtain Environmental Registration for activities to be conducted in an enclosed area;
4   if the GMO contains pesticide features – as do *Bt* varieties – secure a Special Temporary Registry (RET) related to the pesticide legislation;
5   once the RET has been secured, obtain a prior and final technical opinion from the CTNBio before the GMO can be released into the environment;
6   obtain a Temporary Field Experiment Authorization from MAPA;
7   obtain an Operation Licence for Research Areas, required for all GMOs;
8   obtain an Environmental Licence for pre-commercial launching once the field research has been conducted and before the product is commercially launched on the market;
9   procure a further licence, the Environmental Licence for Commercial Launching, once an Environmental Licence is granted;
10 secure CTNBio approval before actual commercialization is initiated (Amâncio and Sampaio, 2005).

To harmonize this regulatory chaos, Bill No. 2401/2003, purporting to establish new rules for regulating biotechnology activities, was passed by the National Congress and approved by the President of the Republic in March 2005. The resulting passage and approval of Biosafety Law No. 11.105 created great expectations among public research institutions, universities, domestic and foreign private companies and venture capital investment funds. In November 2005, Decree No. 5.591 granted CTNBio the power to demand environmental impact studies, if needed, thus reasserting its authority over final clearance of GM products. After CTNBio issues its opinion, the new law requires a second decision concerning social and economic opportunities for commercialization of the GM crop.[13]

Seed companies

The seed market has been estimated at approximately US$1.2 billion, of which 70 per cent is for grain and cotton (Wilkinson and Castelli, 2000). The market is divided among multinational corporations, private domestic companies and state-controlled companies (Campinas Agronomic Institute (IAC) and EMBRAPA). Foreign companies began to enter the Brazilian seed market in the 1960s: Pioneer in 1964, Cargill in 1965, Limagrain and Asgrow in 1971, Dekalb in 1978, and Ciba-Geigy in 1979 (Fonseca et al, 2004). After approval of the new Law on Protection of Cultivars multinational corporations increased their market share in Brazil, taking over several domestic companies.[14] At least 22 Brazilian companies have been acquired by multinational corporations (Fonseca et al, 2004).

Mergers and takeovers have modified the market structure of the three major sectors of the seed industry: hybrids, varieties and vegetables. The hybrids market is the most concentrated and denationalized. The participation of multinational corporations in the maize seed market has been very significant. As can be seen in Table 7.2, the market share of foreign companies jumped from approximately 40 per cent in 1996/1997 to 87 per cent in 2000/2001. EMBRAPA, under a partnership with Unimilho, holds only 5 per cent of the current market.

In the market for variety seeds, the presence of multinational corporations is still rather modest. This segment is dominated by the public sector, especially EMBRAPA, cooperatives and minor regional companies. In 2001, the joint market share of the four major private companies did not exceed 20 per cent of the total output for soybeans, rice and wheat (Martinelli, 2003).

Table 7.2 *Distribution of the hybrid maize seed market in Brazil (1996/1997 and 2000/2001)*

1996/1997		2000/2001	
Companies	**Market share (%)**	**Companies**	**Market share (%)**
Agroceres	26	Monsanto (Dekalb + Agroceres)	48
Cargil	26	Dupont (Pioneer)	13
Pioneer	14	Syngenta	14
Novartis	11	Dow Agro Sciences	7
Braskalb/Dekalb	8	Aventis	5
Dinamilho Carol	3	Agromen	6
Agroeste	1	EMBRAPA/Unimilho	5
Outros	11	Outros	2
Total	100	Total	100

Source: Santini (2002)

In the soybean market, despite increased participation by multinationals from 1996 to 2001, data indicate that EMBRAPA still held the largest share: 46 per cent in 2001. The third position was held by Coodetec (sponsored by the Paraná State cooperative system). Three multinational corporations – Monsoy, Pioneer and Aventis (Bayer) – jointly held only 22 per cent of the market. In the State of Rio Grande do Sul, one of the foremost seed markets, state-controlled companies and cooperatives are still the major suppliers of soybean seeds. During the 2002/2003, state-run companies supplied seeds for 56 per cent of the planted area, while cooperatives provided 18 per cent and multinational corporations 20 per cent.[15]

Commercial production of GM crops in Brazil

In 2005, Brazil was the world's third largest producer of GM crops, with a cultivated area of approximately 9 million hectares (10 per cent of the world total), and had the highest growth rate of cultivated areas, comparing 2004 and 2005 figures (James, 2005). In 2004, approximately one third of the soybean area was cultivated with RR varieties, most of them imported from Argentina.

Nevertheless, Brazil's dissemination of GM crops is way behind that of its world market competitors, the US and Argentina, where such crops began to be cultivated in 1996. Furthermore, RR soybean has been the only GM crop produced in the country, although it is a major player in maize and cotton. In the US, a GM crop variety is produced in six different product groups (pumpkin, cotton, canola, papaya, maize and soybean). Brazil's production is also regionally concentrated in Rio Grande do Sul.

Brazil has lagged behind its competitors mainly because of difficulties in setting up a stable regulatory framework, ending – at least theoretically – in 2005 with the approval of the new biosafety law. However, as mentioned, this did not hinder the clandestine dissemination of GM soybeans, especially in Rio Grande do Sul. Up to 2004, GM soybean seeds were smuggled in from

Table 7.3 *Proportion of soybean growers using GM varieties in four states of Brazil (2004/2005)*

State/region	Conventional	Transgenic*	Total number
Rio Grande do Sul (South)	48,400	81,600 (63%)	130,000
Paraná (South)	116,030	570 (<1%)	116,600
Mato Grosso (Centre-West)	9840	160 (<1%)	10,000
Santa Catarina (South)	9440	560 (<1%)	10,000

Note: *Producers who have executed the 'conduct adjustment commitment' with MAPA.
Source: MAPA (no date)

Table 7.4 *Comparing distribution of farm size in Rio Grande do Sul and 50 largest soybean-growing municipalities*

Farm size	Rio Grande do Sul		Top 50 soybean-growing municipalities	
	Share of farms (%)	Share of area (%)	Share of farms (%)	Share of area (%)
Up to 50ha	86	24	74	26
50–500ha	12	34	22	42
500–2000ha	2	29	3	27
More than 2000ha	<1	13	<1	11

Source: IBGE (1995); calculation by authors

Table 7.5 *GM soy in Brazil, concentrated in the South (2005/2006)*

Regions/state	Planted area (.000 ha)	Yield (kg/ha)	Production (.000 MT)	GM crop share* (100%)
North	500	2770	1400	13
Northeast	1400	2710	3700	20
Southeast	1700	2740	4800	15
South	8100	2390	19,400	61
Rio Grande do Sul	3900	1810	7000	91
Central-west	10,200	2900	29,400	24
Brazil	21,900	2680	58,600	36

Note: MT = million tonnes.
Estimated by Anderson G. Gomes, Céleres Consultants.
Source: Céleres Consultants, www.celeres.com.br

Argentina and Paraguay with negative consequences for the local seed industry, in that state in particular. Furthermore, the GM varieties from Argentina were inadequate for cultivation in the other regions. From 2006, it is expected that Brazilian companies and Monsoy (a Monsanto subsidiary) will offer GM varieties.[16] Table 7.3 shows that until the 2004/2005 crop season, the use of illegal seeds from Argentina was limited to the Southern Region.[17]

Typical GM soybean beneficiaries are growers with 50–500 hectares. Farms of up to 50 hectares account for the largest area in Rio Grande do Sul. (See Table 7.4).

At the opposite extreme, municipalities that specialize in soybean growing have 0.53 per cent of the number of establishments (more than double the percentage for the state) but 10.97 per cent of the area, compared with 12.68 per cent for the state.

Table 7.5 shows that in 2005, following the passage of the biosafety bill and approval of commercial production of RR soybeans, adoption of GM crops increased in several producing regions. However, the quantity of seeds offered on the market was insufficient to ensure adequate planting throughout the country.

Interviews with growers and seed industry representatives in the southern states reveal that the main negative effect of GM soybean introduction (via illegal cultivar imports from Argentina) was a reduction in the percentage of seed acquisition, which is crucial to maintain plant health (see Traxler (1999, 2002) on the critical role of the seed industry in the diffusion of GM seeds in developing countries).

Table 7.6 shows the drastic reduction – from 60 to 5 per cent – in seed use by growers in Rio Grande do Sul due to the introduction of GM cultivars.[18] This trend is all the more striking in light of the fact that, despite the process of concentration seen since 1999, Rio Grande do Sul has 446 accredited seed producers, 581 seed processing units and 42 seed labs (ABRASEM, 2004).

A 'return to normal' depends not only on the Biosafety Act but also on the seed industry's capacity to meet the demand for GM cultivars bearing the characteristics of the elite genetic material available for non-GM cultivars. A 2005 survey showed an estimated 42 GM cultivars[19] (20 Monsoy, 11 EMBRAPA, 7 Pioneer and 4 Coodetec) available for multiplication with the key advantages of adaptation to edapho-climatic conditions and resistance to disease. However, less optimistic estimates reduced the total to nine cultivars, indicating severe constraints.

The institutionalization of GM crops has yet to be consolidated. Indeed, the weakening of the seed industry's most important sector – soybeans – has had continuing negative effects as a consequence of the ambiguity of Brazilian law, which, according to the Biodiversity Convention of 1992 (Dalpoz et al, 2004), allows the multiplication of seeds by farmers themselves. Despite this limitation, GM soybean crops increase every year in regions in which both conventional and GM seed varieties are used, giving rise to persistent criticism by environmentalists, as will be discussed.

Socio-economic issues

Farm income impact assessments

This section summarizes the GM soybean farm income impact assessment by Roessing and Lazzarotto (2005).[20] Use of potential yield by region does not reflect the historical average for local varieties and given levels of management in normal years; GM cultivars obtained by backcrossing rather than plant-breeding programmes initially had yields 3 per cent lower than conventional cultivars, but over time, yields converged as a result of breeding programmes.

The study also analysed differences in physical yield among regions. For short-term GM soybeans, the weighted average for all regions was 2826 kilograms per hectare. For conventional crops and long-term GM crops, the average was 2914 kilograms per hectare.

Seed cost increased by US$20 per hectare (3–5 per cent of crop value). This is more costly and less flexible than charging R$0.60 (US$0.30) per 60 kilogram bag harvested, as was done in the 2004/2005 season under an agreement between cooperatives and Monsanto.[21] Prices ranged from US$5.90 per bushel (very high) to US$5.42 per bushel (close to the current price of R$30 (US$14) per 60 kilograms), reflecting increased production and yields.[22] The study did not consider the possibility of a premium for conventional soybeans.[23]

The overall weighted average for Brazil shows a reduction of 2.9 per cent in the variable cost when compared with conventional soybeans. It is important to note that the variable production cost of GM soybeans is higher than that of conventional soybeans in the centre-west of the country, while the reverse is true in Rio Grande do Sul, where reductions reach 5 per cent of total cost. This explains the success of GM soybeans in that state, even using seeds imported from Argentina. The explanation relates to the key role of glyphosate for weed control in the south, unlike in the centre-west, where many crops are new arrivals and crop rotation is practised far more intensely. This brings us back to the discussion of production scale.

The study by Roessing and Lazzarotto (2005) also shows a cost saving of about 2 per cent per unit of production, with a corresponding increase in the grower's contribution margin, considering payment of US$20 per hectare fees to the company owning the technology (about 5 per cent of total production cost). Again the largest gains are in Rio Grande do Sul, confirming the importance of this type of crop management for growers there and the fact that potential impact projected by the innovation is much less significant in the centre-west.

As shown in Table 7.6, GM soybeans offer a gain of 6 per cent in the medium term and practically no gain at all under current conditions. This raises an important question: if the short-term gain averages null, with payment of technology fees, while medium-term gains result from plant breeding

Table 7.6 *Soybeans: Estimated net income in selected municipalities (US$/ha)*

State/(region)	GM (CP)	GM (MP)	Conventional
Rio Grande do Sul (south)	125	140	117
Paraná (south)	260	280	268
Mato Grosso (centre-west)	126	141	137
Goiás (centre-west)	133	147	140
Brazil (weighted average)	178	195	184

Source: Roessing and Lazzarotto (2005)

programmes designed to insert the gene into strains adapted to different eda-pho-climatic and plant health conditions in different parts of the country, who will undertake the R&D effort and how will it be remunerated?

This highlights the fact that the impact of introducing GM soybeans goes well beyond cost savings. Sharing of gains among stakeholders is crucial. Failing to remunerate the plant breeding labour involved in inserting the gene would greatly undermine the feasibility of GM cultivar diffusion.

Other socio-economic issues

The polarized debate on introducing GM crops in Brazil cannot be understood in the simple framework of 'winners and losers'. A priori, the adoption of a European-type biosafety system is a victory for the environmentalists, aligning Brazil with the precautionary principle. Alternatively, their main goal – to delay the growing of GM crops indefinitely – has been defeated by Law 2401, which says that environmental impact assessments are not mandatory and considers GM crops that do not constitute environmental or health hazards (GM soybeans and cotton had earlier been so classified).

Evidence of the environmental impact of GM soybeans is limited to observations of an increase in the application of the class IV herbicide, glyphosate, which is not especially toxic and, together with Pivot (imidazolinone) and Scorpion, is used by soybean growers instead of highly toxic class I herbicides like Blazer, Flex or Cobra and equally toxic class II products such as Fusiflex, Gamit, Poast and Select.

In terms of the distributional consequences, GM crops favour market-oriented family farmers (Buainain et al, 2002). Furthermore, ultra low volume (ULV) herbicides require more technical knowledge and appropriate equipment on the part of users, so here too the technology favours family farmers, who lack resources to buy equipment for ULV herbicide application.

Table 7.7 highlights two important points: the use of GM cultivars increases seed costs, and, in most regions, reduces expenditure on post-emergence herbicides. The data prove that in 'new' soybean growing areas such as Sinop in the State of Mato Grosso, advantages of using GM soybeans are negligible.

Seed costs, including technology fees, correspond to between 6 per cent and 9 per cent of revenues and 5.4 per cent and 12 per cent of production costs.[24] This shows the importance of the technology fee. The system whereby it is calculated on the basis of the amount of soybeans harvested eliminates the risk of using GM seeds for planting. Paying for seeds transfers additional production and market risks to growers. The fee system clearly points to the limitations of transferring costs associated with the use of modern technology to seeds, which can be multiplied by growers or informal producers (the 'white bag' market in Argentina) as a response not just to additional costs but also to crop expectations. This may indicate a less technology-intensive production strategy.

Table 7.7 *Soybeans: Estimated expenditure on seeds and post-emergence herbicides*

State/(regions)	Seeds (US$/ha)			Post-emerg. herbicides (US$/ha)		
	GM	Conv.	Diff. (%)	GM	Conv.	Diff. (%)
Rio Grande do Sul (south)	52	32	63	14	54	−74
Paraná (south)	51	31	66	12.5	42	−71
Mato Grosso (centre-west)	44	23	92	11.47	25	−53
Goiás (centre-west)	52	32	63	14	19	−24
Brazil (weighted average)	50	29	71	13	35	−55

Source: Roessing and Lazzarotto (2005)

Governance of GM soybean diffusion and adoption is therefore a central issue. A market-centred system in which the seed supplier is an oligopoly with a temporary monopoly on the technology, ranged against a multitude of producers, can lead to negative effects on organization of the industry.

Other factors required to restart the seed industry relate to agreements between growers and Monsanto concerning technology fees and methods of payment. Growers could possibly produce their own seeds as a way of putting pressure for fee reductions on Monsanto and suppliers of genetic material.

It is pertinent to discuss the impact on the small-scale family farming sector. The Landless Workers Movement is clearly opposed to GM diffusion and advocates a total ban. Their position transcends the discussions conducted by FAO (2004) and Silveira and Borges (2004), which were confined to positive and negative technological impacts. Social movements oppose GM technology because they see it as strengthening big agribusiness and concentrating income in the countryside, starting with the obligation for growers to buy seed from oligopolistic suppliers. Specializing in soybean growing does not imply concentration of land tenure; most growers are in the 50–500 hectare group, which is medium size for a farm in Brazil and includes market-oriented family farmers.

What are the benefits of GM soybeans under Brazilian conditions? The conclusions in Roessing and Lazzarotto (2005) largely match those in Pelaez et al (2004):

- yields (production per hectare) are slightly lower;
- variable cost is lower in areas where glyphosate is effective for weed control, as in Rio Grande do Sul;
- fixed cost (not computed in the study) may be reduced if less machinery is used to grow GM soybeans;
- part of the increase in net revenue is offset by the technology fee, ranging from 3–5 per cent of gross sales.[25]

In sum, where is growing GM soybeans advantageous? What led small- to medium-scale growers to defy the law and introduce GM soybeans into Brazil? Here, the contrast with the view taken by Pelaez et al (2004) is striking. Speaking of soybean growing in the US, they conclude, somewhat hastily, that the simplification and greater flexibility of crop management made possible by GM cultivars is not a priority for growers. In the Brazilian case, however, survey data and interviews clearly show that GM cultivars with minimum tillage and two applications of glyphosate recouped crop yields in Rio Grande do Sul.

Identity preservation and market impact

How are GM crops' prospects affected by potential demand for non-GM varieties? There has been little in-depth analysis of market impact, in part because Brazil has not been an 'official' supplier of GM crop varieties. Some countries, notably the EU member states and China, have a demand for non-GM crops and require identity preservation (IP). In Brazil, until 2005 there was a technical protection scheme that guaranteed the existence of large non-GM soybean growing areas, especially in the centre-west. Geographic isolation was and still is responsible for low IP costs, based on documentation and detection. There is no evidence to date that designation of 'free zones' has resulted in market segmentation into GM and non-GM soybeans for export. Brazil's situation is compatible with the hypothesis proposed by Borchgrave et al (2003) and confirmed by interviews in various parts of the country: the non-GM market is based on contracts with buyers who demand IP and pay a premium for it, albeit in the modest range of US\$21–32 per hectare (geared to yields in the centre-west).[26]

Thus, the existence of GM soybeans creates a transaction cost for the non-GM soybean market. If IP segregation and testing at every point in the chain were mandatory for all soybeans produced, growing GM varieties would not be economically viable. Producing GM soybeans would also conflict with market demand in importer countries, especially in the EU, where demand for soybeans as a raw material is far higher than the potential supply of non-GM soybeans covered by 'hard' IP (traceability, segregation and labelling in the final stages). Market segmentation would therefore have different outcomes depending on the institutional regime adopted in countries supplying raw material (soybeans and meal).

Implementation of a complete traceability, segregation and labelling regime as a mandatory requirement for soy exports

Such a regime would require upstream investments along the production chain so that products could be segregated, increasing costs. Interviewee estimates for adapting the entire production system, using appropriate seeds (which would be costly in any situation), isolating crops, changing machinery handling methods and periodically cleaning equipment to avoid contamination range from 4 per cent to 7 per cent of product value, exclusive of testing, certification and inspection costs.

Imposition of such a system would benefit only areas capable of meeting more rigorous detection criteria as they already comply with a milder IP regime. Thus this scenario would be feasible only if entire regions were either GM or non-GM. In the case of Rio Grande do Sul the premium would be lost; it would be assumed that all soybeans were GM crops with lower IP costs. Even so, a wide-reaching agreement would be required to prevent non-GM growers from going to court to ensure compliance with rules prohibiting blends, guaranteed by practices such as crop isolation that increase production costs.

A decision by a producer country to implement the European IP system would paradoxically increase the price and reduce exports of soybeans, as well as encourage production of meal in importer countries. It is hard to predict the effect on the market mix. Non-GM soybeans are currently estimated to account for 20 per cent of the total and their share would probably be increased since, by law, the imposition of IP systems would apply to GM markets as well.

IP regime applied only to non-GM soybeans

This case would produce an interesting paradox: the stricter the IP regime, the higher would be the offer price for soybeans and meal and the lower the price received by growers. The cost of the process could lead to a negative premium, according to several interviewees. In addition, there would be a sharp fall in EU demand for meal, affecting mainly Argentina and the US. This effect would stimulate European production of soy-based protein meal, though exports of non-GM soybeans would not fall significantly.

A survey by Borchgrave et al (2003) highlights the following specific issues for Brazil: the extension of IP to new areas would raise costs by increasing distances between growers and consumer markets, mainly affecting logistics (from US$9 per tonne, or about 5 per cent of product value, to US$15 per tonne); the cost of testing would be in the range of US$3–4 per tonne, or 2–3 per cent higher than the opportunity cost of not planting GM soybeans, which is a little less when the technology fee is included, as shown above.

Thus in more distant areas such as the centre-west, IP costs could reach 12 per cent of product value, a percentage that has never been covered by premiums paid to growers.[27] There is a perception that premiums accrue to trading companies, but this depends on who pays for traceability. The fact remains that even though European consumers prefer non-GM soybean products, higher prices of bulk goods, especially meal, move the demand curve upwards and to the left, reducing prices paid to exporters.

Brazil's situation since 1996 favours establishing a credible process to separate non-GM soy products by contract. This is done by some agribusiness firms for certain applications (mainly vegetable protein for human consumption), to date on no more than 5 per cent of total crushing volumes.[28]

As shown by Borchgrave et al (2003), IP costs mainly affect the early stages of production, creating a problem of how to share them along the chain. In the Brazilian case (Silveira et al, 2006), low IP costs, compared with those of US non-GM growers, reflect the soft IP system adopted on the basis of documentation (tax invoices) and testing only at the port, rather than along the whole production chain. It would be difficult to implement hard IP in Brazil because of deficient infrastructure (i.e. roads, ports and low storage capacity in the countryside).

Conclusion

For seven years Brazil experienced the uncomfortable situation of commercially producing GM crops without recognizing their existence. In March 2005, a framework for legalizing these crops was established with Biosafety Bill No. 11,105. But controversies remain. The 'trench warfare' will no doubt continue to be waged by agencies responsible for health and environmental regulation (ANVISA and IBAMA), the Ministry of Environment, consumer advocacy groups and other NGOs.

At the centre of the debate is the acceptance of CTNBio's competence in managing biosafety. This committee includes representatives of the ministries of justice and agriculture. For environmentalists, each and every GM crop must undergo environmental impact assessment testing and licensing, which would substantially raise the cost of releasing new cultivars. The current law leaves CTNBio free to decide whether such studies and tests are necessary.

The entire debate has revolved around GM soybeans, owing to the fact that growers in the south adopted GM cultivars from Argentina. However, this experience does not necessarily provide the parameters for other crops of relatively small-scale production and economic importance, such as kidney beans or papaya.

Adoption of GM soybeans has produced several lessons in its early stages. The importance of agriculture and agribusiness for the trade balance has a major impact on technology regulation regimes. Countries that import soybeans strongly depend on them as a source of protein. While this favours adoption of IP systems, given the product's low demand elasticity, it also prevents implementation of a moratorium on marketing GM food. Thus, even with the Cartagena Protocol, it is hard to force countries that produce GM cultivars to comply with rules that unify the two market segments. The stricter the rules, the more they increase the cost of non-GM soybeans and the smaller the premium for segregation.

Thus there will not be strong economic incentives to justify investment in infrastructure in producer countries to meet IP requirements. Such investment will be undertaken for sanitary reasons or because governments deliberately intend to implement agricultural traceability systems. The difficulty of doing so in the meat industry production chain shows how unrealistic it is to propose making traceability mandatory throughout the soy chain within a short period of time.

Are other GM crops in the pipeline in the same situation as GM soybeans? Apparently not. GM cotton produced by big growers could be in a similar position, but with the key difference that deciding whether or not to use *Bt* cotton is not subject to a prior requirement of defining GM and non-GM acreages. The advantage of GM cotton is that it offers an additional choice for pest control without forcing growers to take a calculated risk. Given the absence of scale economies in the IP process, it is not easy to see at what point the use of GM crops would be justified in applications where flexibility is fundamental.

It makes no sense to prepare a whole infrastructure to identify GM maize when it is known that its cultivars will be used only in areas and years with expected higher rates of pest occurrence. No farmer would incur the 10 per cent additional variable cost for use of GM seeds without the prospect of higher profit. Unlike GM soybeans, *Bt* technology offers volatile return on investment. One way to mitigate the risk is to introduce another resistance or tolerance gene into a plant without incurring an increase in the technology fee. This matter is still pending in Brazil.

GM diffusion requires acceptance of the idea that all cultivars will have some standard genetically modified traits in the years ahead and that this trend is inevitable. In line with this vision, rigorous prior selection of events by companies, researchers and biosafety committees should guarantee acceptable levels of impact on the environment and human health, based on cost/benefit calculations of using GM cultivars.

Brazil has developed substantial infrastructure and training for agricultural biotechnology research. It has temporary advantages in plant breeding in key research institutions, with EMBRAPA as the most important node. It has prepared institutions for the new intellectual property regime called for by TRIPS (Dalpoz et al, 2004) and built the foundations for advanced research in plant biotechnology, financing networks of genomics and proteomics. It is also prepared to establish collaborative networks with leading global companies and universities, especially in tropical agriculture. But all this effort is insufficient to enable the country to remain a player because other countries, especially the US (including its major corporations), invest far larger sums in research than Brazil, where most of the money for R&D is public.

Thus, if the option to compete in markets for platform products – soybeans, maize, cotton, and some horticultural seeds – becomes less viable as the GM market is consolidated, the alternatives will be smaller, more segmented markets such as those for kidney beans, papaya and banana, and large markets for tropical products such as sugar cane and oranges.

However, this option depends on how Brazil develops its biosafety and hazard warning systems. As noted above, the cost of IP accounts for a relatively small share of value added and does not vary with production scale. By contrast, the return on investment in plant breeding, estimated by interviewees at US$5 million per cultivar obtained over a 12-year period, crucially depends on the scale of the seed market and commercial production, not to mention remuneration of prospecting for the genes involved. Strict requirements for hazard monitoring could make adopting GM crops unviable because, as already mentioned, the investment returns are volatile, especially in relation to pest and disease resistance.

Two basic scenarios can be depicted for the coming years: first, acceptance of GM food by consumers leading to the implanting of several genes in the same cultivar without incurring non-linear cost increases with IP; and second, strict rules on the sale of GM crops, leading to reduced areas under production and a self-fulfilling prophecy in which innovation serves only major markets and products, losing the social function it would have if it served less economically important products normally grown by small farmers. An intermediate scenario would also be possible – segmentation of markets and regions, more tolerance within the country, and a narrower margin for exports to countries with more restrictions. This scenario would depend on the level of strictness.

The main conclusion is that the economic impact of first-generation GM crops does not justify investment in market segregation. Furthermore, charging fees for intellectual property by innovators poses obstacles such as the need to negotiate with growers who are territorially dispersed but organized in associations and cooperatives, as well as limits imposed by the nature of the innovation vector itself, the seed. These general constraints are more severe in countries with persistent technological and socio-economic inequality in the countryside such as Brazil and Argentina (Buainain et al, 2002).

Lastly, the Brazilian case also illustrates the importance of institutional questions and of building a viable and efficient regulatory apparatus. The illegality experienced in Rio Grande do Sul, due partly to Brazil's characteristic tolerance and partly to inept application of the precautionary principle, has had perverse secondary effects, especially on the seed industry in the state, highlighting the importance of information and the danger of the trench warfare waged by environmentalists and their associates.

Notes

1   There is much experience in the use of less sophisticated biotechnology in agriculture. According to Júdice (2004), out of 304 biotechnology companies, 37 are active in the agribusiness sector, operating in fields other than molecular biology-based biotechnology as, for instance, culture tissue for production of seedlings and matrixes, inoculants, and kits for diagnosis of plant diseases, vaccines and biological control vectors.

2   The Centre of Applied Biology at EMBRAPA Maize and Sorghum is conducting field tests of a genetically modified maize cultivar with improved nutritional features. With the use of bioballistics in the PROM regions, involving genes promoting ORF (Parapox virus) resistance, where manipulated, the production of the nutritional protein δ-zein, rich in essential amino acids, was quadrupled. The research team was funded by PRONEX/CNPq (Fonseca et al, 2004).

3 A proposal financed by the Project Finance Agency (FINEP).

4 The Forest Eucalyptus Genome Sequencing Project Consortium, coordinated by FAPESP, involves a consortium of four pulp, paper and wood agglomerate manufacturers: Votorantim, Ripasa, Suzano and Duratex. This study is also of major interest to private companies as it will lead to greater efficiency in wood production and better product quality. The first stage, sequencing, now allows for the identification of the genes responsible for commercially relevant plant features.

5 Waiting for approval to commercial release: (a) *Bt* maize and (b) glyphosate tolerance, by Bayer, Monsanto, Basf, and Syngenta; (c) Rice: Libertlink, to glufosinate tolerance, by Bayer. For the purpose of planned release (research): (a) Tobacco, for medical research, by Fiocruz; (b) Rice, high Lysine, by Monsanto; (c) Lemon Trees, by Allelyx; (d) Maize, by Dow; (e) Cotton, by Dow; (f) Eucalyptus, change in lignin (IN03) by International Paper; Suzano Celulose Allelyx; (g) Sugar Cane, to improve sucrose levels, by CTS; (h) soybeans, tolerant to glufosinate, by Basf, EMBRAPA; (i) Beans, tolerance to golden mosaic disease, by EMBRAPA.

6 CNPq provides support to scientific and technological research through scholarship and research grants, and by funding Regular Lines, Special Programmes and Technological Innovation Programmes. In 2003, 45 per cent of its total investments were channelled to the three fields of knowledge that comprise life sciences: agronomy and veterinary sciences (7.22 per cent), biology (18.56 per cent) and health (19.45 per cent).

7 FAPESP is linked to the Secretary of Science, Technology, Economic Development and Tourism of the State of São Paulo. Its funds are guaranteed by the Constitution of the State, which ensures it a 1 per cent share of the state's total tax revenue.

8 An amount of US$30 million/year was the initial estimate for CT- Biotechnology activities. Agriculture is one of ten areas financed by the fund.

9 The SNPA includes EMBRAPA and its several units, the State Organizations for Farming and Cattle-Raising (Organizações Estaduais de Pesquisa Agropecuária), universities and federal or state research institutes, as well as other public and private organizations directly or indirectly concerned with agricultural and livestock research. Approximately 22 state organizations throughout the five major regions take part in the SNPA. Of these, seven conduct agricultural biotechnology research, including private agribusiness research units like Coodetec (grain cooperatives), CTC (sugar cane cooperatives and sugar mills), CEPLAC (cocoa) and FUNDECITRUS (orange juice coalition).

10 This is a crucial tool for the development of biotechnology companies since it is responsible for the transfer of knowledge from laboratories to the production line.

11 Genomic research projects conducted in Brazil revealed the lack of professionals specialized in this field, particularly as the demand for such professionals increased over time.

12 The Brazilian DPI system includes a Technology Transfer Service, protected by international copyright laws. It provides mechanisms for the dissemination, distribution and transfer of knowledge, e.g. by means of field days, courses, specialist systems, etc. This service is supported by a high-speed Communications Network being set up by a consortium involving FAPESP, EMBRAPA and the World Bank. It also provides, without charge to farmers, a site containing agricultural software (*Agrosoft*), which includes programmes for control and management of rural properties, plague control etc. These programmes are protected by international software laws (Amâncio and Sampaio, 2005).

13 A Council with a board involving 11 ministries is in charge of strategic issues such as appropriate implementation of the Biosafety Protocol (part of the Cartagena Protocol) and its potential impacts.

14 In the late 1990s, Monsanto (US) purchased 29 seed companies, of which 4 are in Brazil; DuPont (US) took over 5, 1 in Brazil; Novartis (Switzerland), 16; Aventis (France/Germany), 9, 4 in Brazil; and Dow AgroScience (US), 13, 5 in Brazil. Sakata Seed Crop (Japan) and Savia S. A. (Mexico) took over the control of 31 companies, of which 3 are in Brazil (Castro et al, 2005).

15 According to Apasul, EMBRAPA had 39 per cent of the total area planted and Monsoy (a branch of Monsanto), 20 per cent. Coodetec varieties reached 15 per cent and a local research Centre, Fundacep, 6 per cent.

16 Issues involving the Cartagena Protocol may interfere with the decision of producers not to grow non-GM varieties in regions with producers of GM cultivars, and vice-versa (see Silveira et al, 2006).

17 The State Government of Paraná prohibits cargoes containing GM soybeans from using the Port of Paranaguá, the second largest Brazilian port for bulk exports. Thus, the spread of GM cultivars in this state is restricted due to political interference, not only due to the inadaptability of Argentine cultivars to the climate and photoperiodism of Paraná.

18 Provisional Measure 131, issued in 2003 to regulate GM soybean growing and marketing, rightly required growers to declare their intention to produce transgenic soybean.

19 Containing Monsanto's Gene 40-3-2 for herbicide tolerance.

20 The EIA was specially prepared for a series of Biotechnology Prospecting Workshops organized by the Strategic Studies Management Centre, an agency of Brazil's Science and Technology Ministry.

21 The areas visited, chosen as representative of the relative importance of soybean growing in Brazil, ranged from Tupanciretã, Rio Grande do Sul and Sinop to Sorriso and Mato Grosso. For details, see Roessing and Lazzarotto (2005).

22 According to Roessing and Lazzarotto (2005), based on data from USDA and Conab, between 1993 and 2004 Brazil's production increased at a geometric rate of 6.28 per cent per annum and yield per hectare rose 3.40 per cent. Argentina, China and India increased production by expanding

acreage, with far more modest gains in yield (0.5, 2.3 and 0.68 per cent, respectively).

23 The premium for non-GM soybeans and meal is estimated at 4–10 per cent of the price, as determined by contract and not by the commodities market.

24 Considering a price of US$5.25 per bushel (60 lbs).

25 Assuming US$5.25 per bushel (60 lbs) and yields of 2800–3000 kg/ha.

26 The barriers imposed by the Paraná state government on exports of GM soybeans from Mato Grosso do Sul relate to the alleged difficulty of guaranteeing segregation at the Port of Paranaguá. The state government was evidently counting on an undeclared moratorium on GM soybeans so that non-GM products could be exported with minimal IP costs. The facts have shown that this option, which contradicts the hypothesis of Borchgrave et al (2003), is unrealistic.

27 According to interviewees, premiums for non-GM soybeans are highest in Japan (around US$7–12 per tonne).

28 The difficulty of implementing a system to segregate beans and meal in Brazil is evidenced by the fact that Bunge, a crusher located in Rio Grande do Sul, buys non-GM soybeans in the centre-west, almost 2000 km away from its mill. It would be hard to segregate, as that would involve a large number of agents who are not willing to change their operating routines.

References

ABRASEM (Associação Brasileira de Sementes e Mudas) (2004) *Anuário*, ABRASEM, Brasilia

Amâncio, M. C. and Sampaio, M. J. (2005) 'Legislação de Biossegurança no Brasil: Cenário Atual, CIB', *Apresentações Técnicas*, CIB (Apresentações Técnica), São Paulo, available at www.cib.org.br

Batalha, M. O., Bonacelli, M. B., Silva, V. M. M. and Borras, M. A. (2004) 'Pós-graduação e biotecnologia: Formação e capacitação de recursos humanos no Brasil', in Silveira, J. M. F. J., Dalpoz, M. E. and Assad, A. L. (eds) *Biotecnologia e Recursos Genéticos: Desafios e Oportunidades para o Brasil*, Instituto de Economia/FINEP, Campinas, pp281–309

Borchgrave, R., Kalaitzandonakes, N., Galvao Gomes, A. and de Frahan, H. (2003) *Economics of Non-GM Food/Feed Supply Chains in Europe*, AgraEurope, London

Buainain, A. M., Souza Filho, H. M. and Silveira, J. M. F. J.(2002) 'Inovação tecnológica na agricultura e agricultura familiar', Wilkinson, J.;and de A. de Lima, D. (eds) *Inovação nas Tradições da Agricultura Familiar*, vol 1, pp47–85, CNPQ-Parelelo 15, Brazil

Castro, A. M. G., Lopes, M. A. and Lima, S. M. V. (2005) 'O futuro do melhoramento genético vegetal no Brasil', EMBRAPA, mimeo

Dalpoz, M. E., Fonseca, M. G. D. and Silveira, J. M. F. J. (2004) 'Políticas governamentais de apoio à pesquisa genômica', in Silveira, J. M. F. J., Dalpoz, M. E. and Assad, A. L. (eds) *Biotecnologia e Recursos Genéticos: Desafios e Oportunidades Para o Brasil*, Instituto de Economia/FINEP, Campinas

EMBRAPA (2004) 'nota informativa: pesquisa biotecnológica na EMBRAPA', EMBRAPA, Brazil

FAO (Food and Agricultural Organization of the United Nations) (2004) 'The state of food and agriculture, 2003–04. Agriculture biotechnology: meeting the needs of the poor?', FAO, Rome HTTP://WWW.FAO.ORG/ DOCUMENTS

Fonseca, M. G. D., Dalpoz, M. E. and Silveira, J. M. F. J. (2004) 'Biotecnologia vegetal e produtos afins: Sementes, mudas e inculantes', in Silveira, J. M. F. J., Dalpoz, M. E. and Assad, A. L. (eds) *Biotecnologia e Recursos Genéticos: Desafios e Oportunidades para o Brasil*, Instituto de Economia/FINEP, Campinas, pp165–199

Fuck, M. P. (2005) 'Funções públicas e Arranjos Institucionais: o papel da embrapa na organização da pesquisa de soja e milho híbrido no Brasil', Instituto de Geociências/Unicamp, Campinas

Furtado, A. T., Silveira, J. M. F. J., Bonacelli, M. B., Salles Filho, S., Zackiewicz, M., Hasegawa, M., Dalpoz, M. E., Bin, A., Diógenes, F., Paulino, S. and Valle, M. (2004) 'Políticas públicas para a inovação tecnológica na agricultura do estado de São Paulo: Métodos para avaliação de impactos da pesquisa', GEOPI, Instituto de Goeciências/Unicamp, FAPESP, Campinas, mimeo

IBGE (Instituto Brasileira de Geografia e Estatística) (1995) 'Censo agropecuário', IBGE, Brasilia

James, C. (2005) 'Preview: Global status of commercialized transgenic crops: 2004', ISAAA Briefs, no 32, Ithaca, NY

Júdice, V. M. M. (2004) 'Biotecnologia e bioindústria no Brasil: Evolução e modelos empresariais', in Silveira, J. M. F. J., Dalpoz, M. E. and Assad, A. L. (eds) *Biotecnologia e Recursos Genéticos: Desafios e Oportunidades para o Brasil*, Instituto de Economia/FINEP, Campinas, pp66–99

Lima, S. M. V. (2005) 'Projeto quo vadis. O futuro da pesquisa agropecuária Brasileira', EMBRAPA Informação Tecnológica, Brazil

MAPA (Ministério da Agricultura, Pecuária e do Abastecimento) (no date) *Estatísticas: Agricultura Brasileira em Números*, available at www.agricultura.gov. br/

Martinelli, O. (2003) *Relatório Setorial: Sementes*, Relatório Parcial da Pesquisa Diretório da Pesquisa Privada, FINEP, Brazil

MCT/CNPq (2004) 'Resenha estatística do CNPq', CNPq, Brazil

Monteiro, A. J. L. C. Lei de (2003) 'Biossegurança: A legislação que não deixam aplicar', CIB, Apresentações Técnicas, São Paulo

Pelaez, V., Albergoni, L. and Guerra, M. P. (2004) 'Soja transgênica versus soja convencional: Uma análise comparativa de custos e benefícios', *Cadernos de Ciência & Tecnologia*, vol 21 (2), pp279–309

Roessing, A. and Lazzarotto, J. J. (2005) 'Soja transgênica no Brasil: Situa-ção atual e perspectivas para os próximos anos', seminar paper, Centro de Gestão e Estudos Estratégicos, Brazil

Salles-Filho, S. L. M. (2001) *Instrumentos de Apoio à Definição de Políticas em Biotecnologia*, FINEP/MCT, Brazil

Santini, G. (2002) 'A reestruturação da indústria de sementes no Brasil', Departamento de Engenharia de Produção/UFSCar, São Carlos

Silveira, J. M. F. J. and Borges, I. C. (2004) 'Um panorama da biotecnologia moderna', in Silveira, J. M. F. J., Dalpoz, M. E. and Assad, A. L. (eds) *Biotecnologia e Recursos Genéticos: Desafios e Oportunidades para o Brasil*, Instituto de Economia/FINEP, Campinas, pp17–31

Silveira, J. M. F. J., Borges, I. C., Vieira Filho, J. E. and Pereira, A. M. (2006) 'The cost of implementing the biosafety protocol: A look at Brazil', IPC Technology Issue Brief, no 1, NEA/IE/Unicamp, Campinas

Traxler, G. (1999) 'Assessing the prospects for the transfer of genetically modified crop varieties to developing countries', *AgBioForum*, vol 2 (3 and 4) (summer/fall)

Traxler, G. (2002) 'Assessing the benefits of plant biotechnology in Latin America', Presentation to IDB

Wilkinson, J. and Castelli, P. G. (2000) *A Transnacionalização da Indústria de Sementes no Brasil: Biotecnologias, Patentes e Biodiversidade*, ActionAid Brasil, Rio de Janeiro

8

China: Emerging Public Sector Model for GM Crop Development

Jikun Huang, Ruifa Hu,
Scott Rozelle and Carl Pray

Introduction

China is often cited as one of the most successful countries in making use of modern biotechnology to improve its agricultural sector (James, 2004; Paarlberg, 2001). However, there has been little work on identifying the policy constraints to further development of plant biotechnology in China. Experts (Huang and Wang, 2003; Keeley, 2003; Huang et al, 2002a) have addressed the record of China's biotechnology research investment but have only touched on biosafety regulations. Almost no one has written about whether the country has experienced or will face serious constraints in these areas, nor examined the even more interesting question: what constraints are holding (or may continue to hold) China back from commercializing additional GM crop varieties?

It is particularly important to understand what can make the regulatory system function effectively enough to ensure that GM products are safe. Opponents, such as Greenpeace, claim that China does not pay enough attention to the possible environmental consequences of GM crop varieties or consumer concerns about food safety. The allegation that unapproved GM rice varieties from pre-production trials have been sold in many retail markets has made international headlines. Some scholars have argued that economic costs have not been considered, such as negative consequences for Chinese agricultural exports. Others fear that multinational life science corporations could dominate China's programme. China's experience may aid other countries in making their agricultural biotechnology sectors function more effectively, and in addressing common institutional challenges associated with the development, extension and management of GM crop varieties.

Several issues need to be examined: how has China developed its agricultural biotechnology programmes? What major policies and institutions have facilitated their development? What have been the social and economic impacts of GM crop varieties? What institutional challenges have policy makers faced and what challenges do they continue to confront? This paper seeks to address these questions by reviewing the experiences and lessons of the first two decades of China's GM crop variety development, and by discussing what has influenced and may continue to affect progress.

Agricultural biotechnology development and policy

Goals and strategies

In the early 1980s, as China initiated its national biotechnology programme, the most frequently stated goal was to create a modern, market-responsive and internationally competitive R&D programme (MOST, 1990, 2000; SSTC, 1990). Other stated goals included improving the nation's food security, promoting sustainable agricultural development, increasing farmer income, improving the environment and human health, and raising China's competitive position in international agricultural markets.

At the heart of the government's strategy was the establishment of a comprehensive, publicly-financed research system (MOST, 2000; SDPC, 2003), including investments to enhance the human and physical research capacity of the national biotechnology scientific community. The need to create a series of institutions and regulations to ensure the healthy and safe development of technology that could both contribute to human welfare and stimulate commercialization was recognized as well. Leaders also wanted to encourage support to the private sector to promote the downstream commercialization of GM products.

While implementation efforts generally have been positive, suggestions for improvements include closer coordination among the players in R&D, more effective and better enforcement of biosafety regulations, promotion of reforms in the seed industry and changes to policies on IPRs.

GM crop development to date

Over the last two decades, the agricultural biotechnology programme has achieved substantial results. *Bt* cotton varieties were approved initially for commercialization in 1997. According to estimates based on information from government sources and commercial firms, the area under *Bt* cotton rose rapidly between 1997 and 2004, reaching 3.7 million hectares (see Table 8.1). We estimate that in 2004 nearly two-thirds of the cotton area was planted to *Bt* cotton, which was being grown by nearly 7 million Chinese farmers. Cotton

Table 8.1 *Bt cotton adoption in China, 1997–2004*

Year	Cotton area (000ha)		Bt cotton share (%)	Source (%)	
	Total	*Bt* cotton		CAAS	Monsanto
1997	4491	34	1	48	52
1998	4459	261	6	20	80
1999	3726	654	18	16	84
2000	4041	1216	30	22	78
2001	4810	2158	45	33	67
2002	4184	2156	52	40	60
2003	5111	2996	59	50	50
2004	5650	3688	65	61	39

Source: Authors' survey

varieties under cultivation come from two sources: Monsanto, a foreign life science company, and national research programmes under the Chinese Academy of Sciences (CAS).

According to a nationwide Ministry of Agriculture (MOA) survey in 1996, scientists have transferred more than 190 genes to more than 100 types of organisms. By 2001, over 60 types of plants were being transformed by 121 different genes (Peng, 2002). By 2003, regulators had approved 1044 applications for agricultural biosafety assessment, 704 for field trial, environmental release and pre-production and 73 for commercialization. Since then, there has been a relatively high annual increase in the number of cases approved by the biosafety system.

Cotton is China's number one cash crop and *Bt* cotton is its most successful GM crop variety. In the late 1980s, responding to the rising use of pesticides to combat a resistant bollworm population, scientists began to develop insect-resistant GM cotton. Starting with a gene isolated from the bacteria, *Bacillus thuringiensis* (*Bt*), they transferred the modified *Bt* gene into major cotton cultivars, using a novel 'pollen-tube pathway' method. Greenhouse testing began in the early 1990s and *Bt* cotton varieties were approved for commercial use in 1997 – four from China's own publicly-funded laboratories and one from Monsanto.[1] The commercial release of *Bt* cotton triggered China's initial experience with GM crop varieties.

Another important achievement has been the development of GM rice. Since research began in the late 1980s, different GM crop varieties have undergone field and environmental release trials and four have reached pre-production. Among them, GM rice with the Xa21 gene, intended to confer resistance to bacterial blight, was approved for environmental release trials in 1997 and one variety entered pre-production trials in 2001. After several biosafety appraisals, in 2004 the Agricultural GMO Biosafety Committee (GMOBC) recommended commercialization. However, the MOA reversed the approval, requiring more detailed biosafety information.

Significant progress has also been made with *Bt* rice varieties able to control rice stem borers and leaf rollers. A number were approved for environmental release trials in 1997 and 1998 (Zhang et al, 1999; Huang et al, 2005a), with GM Xianyou63 and Kemingdao entering pre-production trials in 2001.[2] Another group of scientists introduced a modified CpTI gene into rice and their product was approved for environmental release trials in 1999. After it was introduced into a hybrid variety of rice, GM II-Youming 86, regulators allowed scientists to begin pre-production trials in 2001.[3] Since 1998, greenhouse trials have been underway for GM rice with herbicide tolerance. Scientists are also reportedly working on rice varieties that can withstand drought and salinity.

Other GM crop varieties approved for field trials (some nearly ready for commercialization) include several cotton lines with resistance to fungal disease (Cheng et al, 1997); wheat varieties resistant to barley yellow dwarf virus; maize varieties resistant to insects (Zhang et al, 1999); poplar trees resistant to Gypsy moth; soybeans resistant to herbicides; and potatoes resistant to bacterial disease and Colorado beetles (MOA, 1999; Li, 2000; NCBED, 2000).

In other areas of plant and animal biotechnology, China has produced several recombinant micro-organisms, such as soybean nodule bacteria, nitrogen-fixing bacteria for rice and corn, and phytase, a product made from recombinant yeasts used as a feed additive (Huang, D., 2002). GM nitrogen-fixing bacteria and phytase have been commercialized since 1999. GM pigs and carp have been produced since 1997 (NCBED, 2000). In 2002, China's scientists announced successful sequencing of the rice genome (Yu et al, 2002).[4]

Benefits from GM crops

Critics often express doubt about the usefulness of GM crops for small-scale farmers in developing countries. China's experience to the contrary is well documented.[5]

Evidence on the effects of *Bt* cotton at farm level is based on surveys that we conducted over three years (Pray et al, 2001 and 2002; Huang et al, 2002a, 2002b, 2003a, 2003b). These showed yields of *Bt* cotton to be about 10 per cent higher in 2001 and increasing over time. Farmers also benefited from using substantially less pesticide, which not only increased incomes, but also improved health. Health impacts are particularly important in China where pesticides, applied with small back-pack sprayers used without protective clothing or masks, are held responsible for increased illness among farmers (Huang et al, 2001). Econometric analysis of the survey results concludes that adoption of *Bt* cotton boosts yields between 8.3 per cent and 9.6 per cent and reduces pesticide use by 71 per cent. Projecting to 2010 (using the Global Trade Analysis Programme, reported fully in Huang et al, 2004), benefits would also include a drop in the supply price of cotton, due to increased production, and lower textile costs for both domestic and export markets.

Information on the impact of insect-resistant GM rice is from our recent survey of such rice undergoing pre-production trials (Huang et al, 2005a). Econometric analysis shows increased supply nationally, leading to a price reduction of 12 per cent by 2010 due to yield increases and reduced costs from lower labour and pesticide inputs.

Development priorities

A review of China's policy documents and budgetary expenditures over the past 20 years shows that while GM crop varieties have been developed for more than 60 types of plants, administrators have given priority to rice, cotton, wheat, maize, soybean, potato and rapeseed (see Table 8.2). This is consistent with farmers' demands and national strategic concerns.

Since the late 1990s, research has focused on the isolation and cloning of new disease- and insect-resistance genes, including those conferring resistance to cotton bollworm (*Bt*, CpTI and others), rice stem borer (*Bt*), rice bacterial blight (Xa22 and Xa24), rice plant hopper, wheat powdery mildew (Pm20), wheat yellow mosaic virus and potato bacterial wilt (cecropin B) (MOA, 1999; NCBED, 2000). Significant progress has also been made in the functional genomics of Arabidopsis and in plant bioreactors, especially in utilizing GM plants to produce oral vaccines (BRI, 2000).

China also has focused on crops with the highest potential gains, especially cotton, covering 5–6 million hectares and accounting for 4 per cent of total crop area. In the late 1980s, cotton bollworm resistance caused farmers to use more pesticides on cotton than on any other field crop, and many times as much as most farmers in the world (Huang et al, 2000). By 1995 per hectare pesticide cost for cotton had reached US$101. Cotton producers paid nearly US$500 million for pesticides annually (Huang et al, 2003b).

Research administrators have likewise given priority to quality improvement and stress tolerance (see Table 8.2), for example resistance to drought in response to water shortages in northern China. Although no stress tolerant GM crop variety has been commercialized, relevant budgetary allocations now nearly equal those for insect or disease traits.[6]

Most interestingly, crops receiving highest priority are those produced by relatively poor farmers (Huang et al, 2003d) who are more likely to produce staple commodities such as food and feed grains, oilseeds and fibre crops.[7] Rice, the most important food crop, over the past three decades has accounted for some 27–29 per cent of total grain-sown area (about 20 per cent of total crop area) and been responsible for 41–45 per cent of total grain production (NSBC, 2003).

Table 8.2 *Research focus of plant biotechnology programmes in China*

Crops/traits	Prioritized areas
Crops	Cotton, rice, wheat, maize, soybean, potato, rapeseed, cabbage, tomato
Traits	
Insect resistance	Cotton bollworm, boll weevil and aphids Rice stem borer Wheat aphids Maize stem borer Soybean moth Potato beetle Poplar gypsy moth
Disease resistance	Rice bacterial blight and blast Cotton fungal disease Cotton yellow dwarf Wheat yellow dwarf and rust Soybean cyst nematode Potato bacterial wilt Rapeseed sclerosis
Stress tolerance	Drought, salinity, cold
Quality improvement	Cotton fibre quality Rice cooking quality Wheat quality Maize quality Corn with phytase or high lysine
Herbicide resistance	Rice, soybean
Functional genomics	Rice, rapeseed Arabidopsis

Source: Huang and Wang (2003)

Building national capacity in R&D

Agricultural biotechnology research programmes and institutions

The '863' National High-Tech Research and Development Plan initiated in 1986 by MOST was the first programme to support a large number of applied research projects in this area. In 1997, '973', the National Basic Sciences Initiative, formulated a plan to support basic research underlying high technology, with an emphasis on life science research, which will ultimately contribute to GM crop variety breakthroughs. MOST launched the Special Foundation of Transgenic Plants Research and Commercialization in 1999. It funds only projects jointly submitted by research institutes and companies and requires significant financial commitment from firms wishing to commercialize a technology. Since the mid-1980s the National Natural Science Foundation of China, the major funding agency for basic research, has also allocated significant budget to GM crop variety development.

A number of other inter-ministry and provincial funding institutions have become important supporters of agricultural biotechnology R&D. Since the late 1990s, MOST and the State Development and Reform Commission have jointly sponsored the Key Science and Engineering Programme, which promotes construction of basic research infrastructure (laboratories and other facilities). Most provinces also have a large-scale programme to support high technology, although these mainly support capacity building in their provincial research institutes and provide funding to match grants from MOST and other national sources. Several important internationally-based collaborative projects were begun during the 1980s (for example, with grants from the Rockefeller Foundation and loans from the World Bank), though external funding has accounted for less than 1 per cent of financial resources.

By 2001, nearly 150 laboratories were working on GM crop variety development in more than 50 research institutes and universities across China. Over the past two decades, China has established 30 national key laboratories (NKLs) of which 12 are working exclusively in this area and three others devote a major part of their activities to it (Huang et al, 2001). A large and expanding number of key biotechnology laboratories and agricultural biotechnology research programmes also exist within China's provincial ministries.

A diverse set of national and provincial agencies manage these programmes. MOST, MOA, CAS, the State Forestry Agency and the Ministry of Education are the major national managerial authorities responsible. Their programmes can be vast and complicated. For example, under MOA there are three research academies – the Chinese Academy of Agricultural Sciences (CAAS), the Chinese Academy of Tropical Agriculture (CATA) and the Chinese Academy of Fisheries (CAFi). In 2001, CAAS established a National Key Facility for Crop Gene Resources and Genetic Improvement to consolidate and coordinate the biotechnology research carried out by 12 of its 37 institutes, two NKLs and five key ministerial laboratories. This facility has evolved into one of the most modern biotechnology research centres among the developing countries. CAFi and CATA also have several large biotechnology laboratories and each hosts an NKL for biotechnology. Other centres of agricultural biotechnology are equally as complicated.[8]

Sub-national research programmes typically have an institutional framework similar to the national one: each province has its own academy of agricultural sciences and at least one agricultural university, while each academy or university has one or more institutes or laboratories focused on agricultural biotechnology. Even at prefecture level, a number of research institutes have been working on applied GM research. The number of these programmes and institutes continues to grow.

Agricultural biotechnology research capacity and investment

Creation of a modern and internationally competitive biotechnology R&D system requires substantial investment in human resources and basic infrastructure. Since the early 1980s, China's public investment in this field has increased significantly, creating a large and growing number of scientists. Data presented in Table 8.3 are based on a recent nationwide survey of agricultural biotechnology research institutes by the Centre for Chinese Agricultural Policy (CCAP).

The number of agricultural biotechnology researchers has grown steadily since the mid-1980s (see Table 8.3). In 2003, about 5770 researchers were working on the subject, of which about 3200 were considered part of their laboratory's professional staff. Nearly 60 per cent of total research staff work on plant biotechnology while 30 per cent and 10 per cent do research on animal and micro-organism biotechnology, respectively.

Earlier CCAP surveys demonstrate that the quality of human resources focused on biotechnology research has improved over time (Huang et al, 2001). The share of PhD researchers grew from only 2 per cent in 1986 to more than 20 per cent in 2000. While still low by international standards, this percentage is much higher than in China's general agricultural research system, where PhD researchers accounted for only 1.1 per cent of total professional staff in 1999 (Huang et al, 2003c).

Even more spectacular growth has occurred in China's agricultural biotechnology research investment (see Table 8.4), estimated at only US$10 million in 1986 when China formally started its 863 Plan, but at US$25 million by1990. During this period the research project budget nearly tripled

Table 8.3 *Research staff in agricultural biotechnology in China, 1986–2003*

Year	Total	Plant	Animal	Micro-organism
All staff				
1986	1436	917	378	142
1990	2189	1271	691	227
1995	2918	1750	873	295
2000	4502	2545	1476	481
2003	5772	3125	1901	746
Professional staff				
1986	667	391	217	60
1990	1060	615	335	110
1995	1561	936	467	158
2000	2470	1396	809	264
2003	3203	1735	1055	413

Source: Authors' survey

and expenditures on laboratories and equipment practically doubled. While the growth rate slowed between 1990 and 1995 (investment in biotechnology equipment having been largely completed in the early 1990s), in real terms annual growth rate in the research project budget remained as high as 6 per cent during that period.

Accelerating once again during the past decade, China's biotechnology research investment increased from US$33 million in 1995 to US$104 million in 2000, representing an annual growth rate of some 26 per cent. From 2000 to 2003, the budget further doubled (see Table 8.4). In 2003, China spent nearly US$200 million (or US$953 million in PPP terms) on agricultural biotechnology, US$120 million of it on plant biotechnology.

While media and international reports sometimes highlight the role of foreign life science firms and the domestic private sector, investment in China's biotechnology is overwhelmingly from government sources. According to our survey, public investment accounted for 97 per cent of the agricultural

Table 8.4 *Public expenditures on agricultural biotechnology in China, 1986–2003*

Year	Total	Plant	Animal	Micro-organism
Million RMB in real 2003 prices				
1986	89	51	28	10
1990	204	118	64	23
1995	273	157	86	30
2000	861	450	323	88
2003	1647	996	468	183
Million US$ converted at official exchange rates				
1986	10	6	3	1
1990	25	15	8	3
1995	33	19	10	4
2000	104	54	39	11
2003	199	120	57	22
Million US$ converted at PPP				
2003	953	576	271	106

Note: Expenditures include both project grants and costs related to equipment and buildings. Official exchange rate in the corresponding year is used to convert the domestic currency to US$ at current price. Official change rate was 8.277 RMB/US$ and the converting rate of RMB to US$ in PPP was 1.729 in 2003.

Source: Authors' survey

biotechnology budgets of the 46 biotech research institutes in 2003. Budgets from publicly-funded competitive research grants accounted for two thirds of the total. This share has increased over time, showing that China's biotechnology development is moving from the capacity building to the research stage.

Recent interviews with MOST officials and research administrators confirm that the Ministry continues to accelerate its investments. The goal is to make China one of the world's premiere centres for plant biotechnology research – an ambition embodied in the country's 11th Five Year Plan (2006–2010).

Remaining challenges

Coordination and consolidation will be essential to make the research programme more effective and internationally competitive. While research capacity has improved significantly in both quantity and quality, human resources may still need further development. China must also set clear priorities in order to maximize benefits from investments and meet strategic goals. Until now, agricultural biotechnology research investments have emphasized technologies geared to meeting food security objectives and farmers' demands for reducing pesticides and raising productivity and competitiveness. Research administrators are also investing heavily in stress tolerance, with potential implications for both food security and poverty alleviation.

But some strategic questions remain open: should China continue investing only its own resources in biotechnology or should it encourage investment by foreign life science firms? What should be the role of the domestic private sector? Is it necessary to continue expanding biotechnology at the sub-national level or should it consolidate its resources? How can biotechnology programmes at different levels (and by different actors) be coordinated to maximize the efficiency of the nation's research investments?

Agricultural GM product biosafety regulation

Institutional setting

After nearly a decade of experience, China has developed a comprehensive biosafety regulatory system. The Joint Ministerial Meeting (JMM) is in charge of directing and creating a comprehensive policy on regulation. Established by the State Council (the highest governmental body) through the issuance of the 'Regulation on the Safety Administration of Agricultural Transgenic Organisms', the JMM is composed of high-level representatives from MOA, MOST, the National Development and Reform Commission, the Ministry of Health (MOH), the Ministry of Commerce, the National Inspection and Quarantine Agency and the National Environmental Protection Authority. The group is responsible for coordinating key issues related to agricultural GM product

biosafety and examining, approving and setting up all policies and major regulations concerning agricultural commercialization, labelling, imports and
exports.

The MOA is the primary institution in charge of implementing agricultural biosafety regulations and GM product commercialization. Within MOA, the
Leading Group on Agricultural GM Product Biosafety Management oversees
the work of the GM product Biosafety Management Office (BMO) while scientific assessment is the responsibility of the National Agricultural GMOBC.[9]
This committee meets twice a year to evaluate all biosafety applications related
to experimental research including field trials, environmental release trials,
pre-production trials and commercialization. Partly because of shared respon-
sibilities, there is a good deal of uncertainty in the decision-making process
concerning high-profile technologies or experiments on which MOA/Leading
Group members do not have a clear view. This can lead to substantial delays
in getting final approval.

The MOH is responsible for the food safety management of biotechnology
products and submits technical issues to an Appraisal Committee of experts
and scientists in food, health, nutrition and toxicology. Acting on the Committee's decisions, the MOH issues formal assessments on the food safety properties of GM products and grants food safety certificates for novel ones. During
the process of assessing the overall biosafety of each GM product experiment
and commercialization application the certificate must be submitted to BMO
and GMOBC by scientists/research administrators.

The State Environmental Protection Authority also participates in GM
crop biosafety management through the JMM, having assumed responsibility
for managing and implementing international biosafety protocols. Its primary
domestic biosafety duties are to understand and report on the effect of the
release of GM crops.

Also China has been trying to set up a local network of GM product biosafety regulation and management offices. By 2005, 26 provinces with agricultural biotechnology R&D programmes had established provincial-level
agricultural GM crop biosafety management offices to monitor and inspect
the performance of research, field trials and the results of decisions allowing
the commercialization of agricultural biotechnology products in their provinces. Their primary role is to monitor the local implementation of biosafety
regulations.

Biosafety regulations

With the continued development of agricultural biotechnology, rising GM
product imports and concerns about consumer health and perceptions of
food safety, China has periodically amended its biosafety regulations. In 1993
MOST issued 'Measures for Safe Administration of Genetic Engineering', the
first comprehensive set of biosafety guidelines (see Table 8.5). MOST then
required relevant ministries to issue their own, more detailed, sets of biosafety

regulations. In 1996 MOA issued 'Implementation Measures for Safety Control of Agricultural Organism Biological Engineering', its first set of biosafety regulations covering plants, animals and micro-organisms (see Table 8.5). One subset spelled out procedures that research institutes, agencies and firms would need to follow to meet minimum biosafety standards at each stage of GM product development. In the mid-1990s, separate biosafety standards were set up for small-scale field trials, environmental release trials and commercialization. In 1997, MOA officials established the first Biosafety Committee to provide the Ministry with expert advice on biosafety assessment.

The State Council replaced MOA's 1966 guidelines with a new set of regulations covering plants, animals and micro-organisms in May 2001 (see Table 8.5). These included a new stage of assessment for GM products. In addition to small-scale field trials and medium-scale environmental release trials, scientists were required to successfully put their GM crop varieties through pre-production trials. That meant that a scientist had to pass through three stages of approval to commercialize a product. The State Council's new regulations also dealt with processed food products, initially not covered. Labelling became required on all products that contained GM food components.[10] In response to rising soybean imports and prospects for maize imports, new regulations governed the export and import of GM crop varieties and any food products containing them. The national guidelines also pressured provincial-level and sub-provincial agencies to implement stricter monitoring guidelines.

In April 2002, MOH promulgated its first comprehensive set of regulations on GM food hygiene, requiring, among other conditions, that all fresh and processed food with a GM crop component be approved by MOH before being sold in any market. These also duplicated the State Council's call for the labelling of GM foods.

MOA released a more comprehensive set of biosafety regulations in 2004 in response to the proliferation of unapproved (and illegal) *Bt* cotton varieties in northern China and the Yangtse River Basin. These included the requirement that any newly developed GM variety already have an MOA safety certificate before being allowed to enter regional variety trials (the last stage in the process through which crop varieties are certified for commercialization). In order to have a new cotton variety certified applicants have to present a certificate specifying whether a variety is GM or non-GM, in addition to providing yield and other performance information.

At the same time, MOA sought to simplify the certification and approval administrative process, stating that anyone with a *Bt* cotton variety that has received production safety or commercialization certification from one province can apply directly for safety certification from another province in the same cotton-ecological region (of which there are three in China).[11] Also, developers of a variety can apply directly for safety certification in one province if that variety was developed from a parent variety that already has been granted safety certification. The impact of this streamlining was significant. In 2004 the number of varieties for which scientists received biosafety certificates soared: about

Table 8.5 *Major policy measures related to agricultural biosafety regulation in China*

The MOST Biosafety regulation in 1993	MOST issued 'Measures for the Safety Administration of Genetic Engineering' in December 1993. It includes biosafety categories and safety assessments, application and approval procedures, safety control measures and legal regulations. These measures provide a framework for each sub-sector's development of its own detailed regulations and implementation procedures.
The first MOA agriculture biosafety management regulation in 1996	Based on MOST's 1993 Measures, MOA issued 'Implementation Measures for the Safety Control of Agricultural Organism Genetic Engineering' in July 1996. It covers plants, animals and micro-organisms.
Agriculture GMO Biosafety Committee in 1997	Ministry level Agricultural GMO Biosafety Committee was set up in MOA in 1997 and updated to national level with its office in MOA in 2002.
The State Council agriculture biosafety regulation in 2001	The State Council amended MOA's 1996 agricultural biosafety regulation to include trade and labelling of GM farm products and issued the new 'Regulation on the Safety Administration of Agricultural Transgenic Organisms', effective as of 23 May 2001.
The second MOA agriculture biosafety management regulation in 2002	Based on the State Council's 2001 Regulation, MOA amended its 1996 biosafety implementation regulation and issued three independent ones, effective as of 20 March 2002. They are 'Measures for Management of the Evaluation on the Safety of Agricultural Transgenic Organisms', 'Measures on Safety Administration of Agricultural Transgenic Organism Imports', and 'Measures on the Labelling Administration of Agricultural Transgenic Organisms'.
MOH regulation on GM product food hygiene in 2002	In April 2002, the MOH issued a management regulation on GM product food hygiene, which requires that all foods, including processed foods, using GM crops as materials must be approved by MOH before being sold in the markets and that GM crop foods must be labelled.
MOA regulation on regional variety testing of GM crops in 2004	In July 2004, MOA issued a policy to enhance the management of GM varieties. It requires that any newly developed GM varieties must have safety certificates from MOA when applications are made for regional variety trails. For any cotton varieties, applicants must present a certificate attesting that it is a GM or non-GM variety, obtainable through MOA-designated testing institutes/organizations that conduct gene tests.

| Regulation on GM commercialization scale based on ecological region in 2004 | In September 2004, MOA issued policies to simplify regulation procedures on *bt* cotton's commercialization: (1) anyone receiving *bt* cotton variety (or variety line) production safety (or commercialization) certification from one province can directly apply for safety certification in any other province within the same cotton-ecological region (of which China has three); (2) anyone receiving *Bt* cotton variety (or variety line) production safety (or commercialization) certification from any region can directly apply for safety certification in provinces in other regions; (3) anyone with a variety (or variety line) can apply directly for safety certification in one province if the variety (or variety line) is back-crossed with a variety (or variety line) with safety certification. |

Table 8.6 *Estimated government budgetary allocations on agricultural biosafety research and regulation implementation*

Year	Biosafety research budget (million RMB) (a)	Biosafety administrative budget (million RMB) (b)	Total biosafety budget (million RMB) (c) = a + b	Total biosafety budget (million US$)	As share of biotech research budget (%)
1997	0.45	0.56	1.01	0.12	0.23
1998	0.58	0.60	1.18	0.14	0.20
1999	0.72	0.66	1.38	0.17	0.16
2000	1.71	0.70	2.41	0.29	0.28
2001	8.69	0.77	9.46	1.14	0.82
2002	11.68	3.38	15.06	1.82	1.11
2003	17.92	5.46	23.38	2.83	1.44
2004	18.64	5.52	24.16	2.92	NA

Note: Budgets are deflated by CPI and in 2000 constant prices. The total government expenditure on agricultural biotechnology is from a recent CCAP survey of agricultural biotechnology institutes and administrative agencies.

Source: Huang et al (2005b)

130 new certificates were issued – far more than in 2003. Interestingly, nearly all of these newly approved varieties were already being planted in farmers' fields.

Compared to standards in other developing countries, China's improved biosafety regulations seem fairly comprehensive. But alongside requirements in many developed countries, they are considered incomplete and rather loosely implemented. Many observers believe their effectiveness has been limited.

With regard to GM commodity imports, if a gene has passed through the biosafety regulatory process in the US or Canada, national authorities generally assume that it meets China's food safety and environmental regulations. Additional requirements are imposed only when a foreign technology is imported; it must then be tested for pest and/or disease control efficacy under Chinese field conditions.

When China started commercialization of *Bt* cotton in 1997, we estimated that the total budget allocated to biosafety research and regulatory management was US$120 thousand (see Table 8.6). Nearly half was used for biosafety research on *Bt* cotton lines generated by CAAS. The rest went for regulatory costs of running the Biosafety Committee. There have been substantial government budgetary allocations for research on and management of agricultural biosafety, and for building capacity for regulation of GM product biosafety at both national and provincial levels. Currently about US$3 million is spent annually on agricultural biosafety (excluding expenditures for labelling and market inspection).

Remaining challenges

Despite these efforts, the record of regulating GM events in the field has not been adequate. China has millions of small-scale farmers, thousands of seed companies and hundreds of crop breeding research institutes. After the first five varieties of *Bt*-cotton were approved for commercialization in five provinces in 1997, research institutes in almost all provinces generated a host of their own *Bt*-cotton varieties in their breeding programmes.[12] In many cases, existing (approved) varieties are being used in these new varieties through backcrossing methods. As the demand for *Bt*-cotton seed increased, seed stocks for most of the new varieties were generated and disseminated through conventional seed trial and regulatory systems, completely bypassing the GMOBC and other institutional structures set up to regulate GM crop varieties. In fact, our field surveys reveal that the number of illegal *Bt*-cotton far surpassed the number of legal *Bt*-cotton varieties after only several years of commercialization.

Although the biosafety regulation and implementation systems have improved over time, additional problems have emerged. First, the system monitoring GM crop variety production at local levels is relatively weak. For example, between 1999 and 2001 we surveyed 854 *Bt* cotton farmers who had adopted 28 *Bt* cotton varieties with 73 different varietal names. But of these, only 13 varieties (46 per cent) had been approved for commercialization by the

GMOBC and were registered with the BMO, implying that a large number of illegal *Bt* cotton varieties are being disseminated. It has been shown that while both legal and illegal *Bt* cotton varieties have some tendency to reduce pesticide use, yields of legal *Bt* cotton varieties are higher than those of illegal *Bt* cottons, which only yield the same as non-*Bt* cotton varieties (Hu et al, 2005). Hence, farmers benefit much more from legal *Bt* cotton varieties that have gone through the regulatory system.

Second, a better system is needed for monitoring GM crop variety trials before commercialization. The recent experience of farmers in pre-production trials in Hubei province selling their output in the market shows the absence of effective monitoring. Officials recognize that such incidents hurt the entire GM R&D and commercialization effort by undermining confidence in the system's ability to effectively regulate GM biosafety.

Third, officials understand the need for more transparency in the biosafety evaluation process. The main problem appears to be that criteria on which decisions are based are not always explained to the scientists or research administrators trying to get their new technologies approved.

Fourth, issues remain about who should bear the cost of biosafety regulation, especially who should financially support biosafety assessments throughout the GM product trial process. If a technology is generated by the public sector, who should fund the effort to push it through the biosafety regulation procedures after laboratory research has been completed? With costs of meeting biosafety assessments rising sharply, many provincial and prefectural research institutes find themselves unable to afford them despite having generated lines of GM crops that they consider to be of high quality. Given these constraints, the proliferation of illegal *Bt* cotton in recent years is not surprising – especially prior to the reforms of 2004.

Finally, although investment in agricultural GM crop biosafety regulation has increased significantly, further budgetary increases are required to make the system effective. In 2003, the total public budget allocated to agricultural GM product biosafety regulation was equivalent to only 1.4 per cent of the government's agricultural biotechnology expenditure (see Table 8.6, last column).

Commercial dissemination: Policy shifts and impacts

Intellectual property rights

While no one claims that China has strong intellectual property rights, they have improved remarkably over the past two decades. Prior to the late 1990s neither plant breeders nor seed companies had any legal control over the varieties or genes that they created. It was legal for a seed company to take a new variety developed by another company and reproduce and market it. Nor was there any restriction on the use of that new variety as a parent in the development of another variety. Varieties could be sold and marketed legally without paying any licensing fees or royalties to their creators. Use of other breeders' varieties as parents was a common practice and remains so today.

In 1997, things began to change. The first Plant Variety Protection (PVP) Act was passed and legislators developed a fairly comprehensive legal framework based on the UPOV.[13] Though it is unclear what impact the PVP has had on the revenues of breeding institutes or seed companies – or the incentives of their scientists and administrators – recent research has shown that individuals, firms and institutes are increasingly interested in applying for patents. Using a database of all major varieties of 22 crops, it has been shown that applications for plant variety certificates (PVCs) increased from 115 in 1999 to more than 400 in 2003 (Hu et al, 2005). Although research institutes have accounted for most of the applications (59 per cent), companies and individuals applied for 33 per cent of the PVCs.

Gaps remain in IPR regulations. Most glaringly, the PVP still does not cover cotton. Even if it did, it would not restrict use of commercial varieties as parents in the production of other varieties because a research exemption explicitly allows this. Nor does the PVP protect novel genes developed by scientists.

It is through the patent system that firms try to prevent their proprietary varieties and novel genes from being used by other scientists without permission. Dr Guo Sandui of CAAS received a patent on the *Bt* gene that he developed, which is being used in all of the CAAS varieties (Fang et al, 2001). Monsanto has patents on several genes that are important in the production of GM plant varieties. These patents cover genes that are inserted into plants to make them resistant to certain classes of insects, processes that create GM cotton varieties and genes that promote the expression of the gene to which they are attached.

Firms also use trademarks – another form of intellectual property – to protect their technologies. BioCentury, a company that manages CAAS' *Bt* cotton, has trademark protection on its name and some components of its technology (Fang et al, 2001). Monsanto's Bollgard trademark on its *Bt* cotton varieties prohibits other firms from using that name.

While seed companies in the cotton industry have taken steps to protect their varieties, it is unclear how effective these have been. Interviews with CAAS officials, BioCentury Seeds and international biotech companies in China have shown that current IPR laws and their enforcement provide little protection. We were told that innovative cotton seed firms have had little success in keeping other firms from copying their varieties or trademarks. New entrants in the *Bt* seed market reproduce, backcross and market the varieties developed by both domestic and foreign life science firms/research institutes. We estimate that while the varieties with CAAS's gene covered 2.25 million hectares in 2004 (61 per cent of the national *Bt* cotton area (see Table 8.1)), the *Bt* cotton varieties sold by CAAS and BioCentury Seeds covered only 5 per cent of the total *Bt* cotton area in that year. The manager of BioCentury said that the rate was even lower (only 3 per cent) in 2003 when *Bt* cotton varieties with the CAAS gene covered 1.51 million hectares (50 per cent of the total *Bt* cotton area (Table 8.1)). Only a slightly higher rate was recorded for cotton varieties with Monsanto's *Bt* gene. According to the Delta and Pine Land representative in Beijing (May 13, 2005) legitimate Bollgard seed covered less than 5 per cent of the *Bt* cotton market in 2004, despite the fact that varieties with Monsanto's gene cover at least 39 per cent of the total *Bt* cotton areas.

The seed industry

China's seed industry has evolved over time. In the mid-1990s, state-owned enterprise (SOE) seed monopolies dominated, with 2700 SOEs operating in local counties, prefectures and provinces. In most counties, only the local region's SOE was allowed to sell seeds of the major crops. Regulations banned non-SOE seed firms from participating in the production, distribution and sale of hybrid maize and rice. Typically, county- and prefecture-based SOEs sold their seed through township agricultural extension agents. In 2000, the government passed a new seed law that, for the first time, legally defined a role for the private sector. Among other things, it states that any entrepreneur with access to the required minimum of capital and facilities can sell seed. Private companies are allowed to sell seed bred by public institutes, including all varieties of hybrid maize and GM or non-GM cotton. This legislation nullified the monopoly positions of county, prefectural and provincial seed companies, creating new distribution channels for seeds alongside those of the agricultural extension system. Private firms, quasi-commercialized SOEs and traditional SOEs were all allowed to apply for permits to sell seed in any jurisdiction. Firms were also permitted to have their seeds certified at the provincial level, entitling them to sell seed in any county in the province. For the first time, it became feasible for national companies to establish their own distribution and retail networks.

These changes have allowed a commercial, competitive seed industry to evolve. New firms have entered the market and fresh sources of investment have emerged, including foreign investment (though still low). By late 2001, nine companies had permits to sell seed anywhere in China. Thousands of small seed companies opened up to supply local needs. All this has been particularly important in developing a competitive commercial cotton seed industry, especially in terms of creating and marketing GM cotton varieties.

Policy shifts and the impact on producer efficiency

Our study seeks to answer three questions. First, will stronger IPRs on novel genes and varieties provide greater incentives to do research and promote efficiency in cotton production? Second, will seed industry reforms help or hurt production efficiency? And third, will more effective approval procedures and biosafety regulation enforcement benefit farmers? (More comprehensive discussion of these issues can be found in Hu et al, 2005.)

At the outset, our study required that we determine the sources of farmers' seeds. While respondents could state precisely from which source their seed came, it was more difficult for them to identify the company that had produced their varieties. Farmers buying from seed companies typically could distinguish between 'foreign seed' varieties using the Monsanto gene and those developed with the CAAS gene. However, they were often unsure as to whether the seed they had purchased was legitimate (i.e. produced and distributed by Jidai or BioCentury or their authorized partners and/or dealers). We found that the legitimacy of seed could be determined by the price farmers paid, and by whether or not it was delinted and/or treated. Legitimate Jidai and BioCentury seed is always delinted and treated and nearly always sold at a fixed price of about 40 yuan (US$4.84) per kilogram. So when farmers reported that they bought seed for less than 30 yuan (US$3.63) per kilogram, and when they said that it was loose, fuzzy and pink (showing that it had not been delinted and treated) we assumed that it was illegitimate.

Next, we divided all cotton seed used by farmers into nine types: legitimate Monsanto, Delta and Pine Land (MDP) (high priced, delinted and treated);[14] illegitimate MDP (low-price or fuzzy and non-treated); legitimate CAAS (high priced, delinted and treated); illegitimate CAAS (low-price or fuzzy and non-treated); unapproved seed from a seed company (neither MDP nor CAAS varieties); seed from the Agricultural Extension Station/Cotton Office; seed from a production base (as well as some bought from a ginning factory, both owned by the county's agricultural bureau); and self-saved seed.

Our first exercise was to assess whether proprietary *Bt* varieties have any economic advantage over non-proprietary varieties (legitimate MDP vs illegitimate MDP; legitimate CAAS vs illegitimate CAAS). Second, we sought to analyse the impact of a strong biosafety management system (approved varieties – for example, legitimate MDP or legitimate CAAS vs unapproved varieties). Third, we tested whether or not the seed sold through the market (whether legitimate or not, including seed sold by the seed production base and ginning plant) performed as well as seed sold through traditional non-market channels (agricultural extension station or the Cotton Office). We also tested whether or not the varieties of the multinationals performed as well as domestic ones (legitimate MDP vs legitimate CAAS).

Identifying the differences in efficiency

To understand the net effect of the seeds (and genes) from different sources, we pooled data, for three years, from five provinces. We followed an approach similar to the one used in our impact studies but included the specification of an additional set of seed dummy variables in both pesticide and yield equations. (Hu et al, 2005, fully explore the impacts of various actors on pesticide use and crop yield but in this chapter we pay attention only to the three issues raised.)

The results of our models, empirically estimated, show that the potentially positive effect of IPRs can be analysed by looking at the differences in results obtained using legitimate and illegitimate MDP and CAAS varieties. If IPRs were enforced, the illegitimate seed would not be available and more farmers would be using legitimate seed. When farmers used either MDP or CAAS legitimate seed, pesticide use fell between 39.77 and 41.45 kilograms per hectare. When using the illegitimate MDP or CAAS seed, there was less fall in pesticide use (30.57 and 33.52 kilograms, respectively). If IPR regulations had kept unauthorized varieties from being sold as MDP or CAAS varieties, farmers would have used less pesticide.

The effect of enforcing IPRs is even clearer in yield equations. The yields of the legitimate MDP varieties were 25.7 per cent above those of the baseline conventional varieties, while yields of illegitimate MDP varieties were only 12.8 per cent higher. The same was true in the case of CAAS varieties of *Bt* cotton. Yields of the legitimate variety were 19.2 per cent greater than those of conventional varieties while yields of illegitimate versions of CAAS varieties were not statistically different than those of conventional varieties.

The results also show the possible effect of enforcing biosafety regulations – especially in the case of yields. We can see the benefits of biosafety regulation by comparing the coefficients of the seeds with approved *Bt* genes and unapproved *Bt* genes. The pesticide equation shows little difference in reduction of pesticide use. The seed that did not go through the biosafety regulation process is almost as effective (38.53kg reduction) as that of the legitimate MDP and CAAS varieties (39.77/41.45kg) and statistically indistinguishable. However, the yields of unapproved varieties are not statistically different from those of conventional varieties of cotton, while yields of the approved MDP and CAAS varieties are 19.2–25.7 per cent higher.

Finally, our results suggest that seed market reform has improved the performance of *Bt* cotton varieties. When seed varieties sold through traditional channels are used, there is less reduction in pesticide use and yields are almost all lower than with seed purchased through the market. Hence, all of our results show the importance of promoting a stronger IPR system, a more effective network of biosafety regulatory bodies and a stronger seed industry.

Concluding remarks

China considers agricultural biotechnology as one of its most important, strategic tools for improving national food security, raising agricultural productivity and creating a competitive position for its farmers in international agricultural markets. A wide array of agricultural biotechnology products have been developed and disseminated. Several GM crop varieties are in the pipeline for commercialization that can potentially create high returns for domestic farmers.

Despite this success and further potential at the farm level, China's institutional framework is complex and evolving at both national and local levels. The growth of government investment in agricultural biotechnology research has been remarkable. There has also been significant improvement in the level of human resources working in this field. Examination of the foci of agricultural biotechnology research reveals that food security objectives and farmers' demands for specific traits and crops are being incorporated into the technologies being given priority by the research system. The recent emphasis on developing drought resistant and other stress tolerant GM crops also suggests that biotechnological products are not only being geared at high-potential areas, as critics argue, but also at the needs of poor farmers. Our review of agricultural biotechnology R&D does not show any significant constraints to the national government's ability and willingness to continue funding in this area. However, there are outstanding issues. For example, better coordination among institutions and strategic consolidation of the programmes is needed. Important decisions also have to be made as to whether to encourage (or continue to discourage) the participation of domestic private and foreign partners in the agricultural biotechnology sector.

Our study shows that China is developing one of the most comprehensive agricultural GM crop biosafety regulatory systems in the developing world. Beside the national GM product biosafety regulation system, subnational agricultural biosafety regulatory offices have been established in nearly all provinces. Investment in the biosafety system has increased rapidly in recent years and biosafety regulation has become progressively more comprehensive and sophisticated. Our statistical analysis shows that *Bt* cotton varieties approved by the biosafety committee perform better than varieties that are not approved. This implies that enforcing biosafety regulations could have substantial positive impact on farmers. Results also imply that poor regulations are slowing down the adoption of high quality *Bt* cotton and reducing economic welfare. Costs to farmers of better enforcement of regulations need to be carefully weighed against their future possible benefits, such as pest susceptibility to *Bt*. Our results also show that the enforcement of biosafety regulations can be a substitute for formal intellectual property rights. If the government eliminated the use of unapproved varieties, allowing only approved varieties to be used, this would essentially establish a duopoly for CAAS and MDP.

Our review of agricultural GM product biosafety implementation shows that enforcement is not without problems. The important caveat here is that government investment in biotechnology regulation, although recently rising significantly, will have to be further increased to ensure that regulations are enforced and effectively implemented. Current levels of investment in both biosafety research and regulatory management are not sufficient. Investment should focus on more monitoring of GM crop production and varietal breeding both before and after commercialization. There should be a clear division of – and sufficient funding for – GM product biosafety assessments.

Reviews of IPR legislation and seed industry reforms show that the environment for commercialization has been improving, particularly in recent years. Domestic firms in China and international seed firms have started to invest in the seed industry in response to a new seed law and changes in regulations. These changes, along with the development of the nation's agricultural biosafety regulatory system, have led to greater investments by commercial seed firms in developing and spreading new varieties. However, it is important to note that – even with recent progress – these investments are still small compared to those of OECD countries (Pray and Fuglie, 2001).

China's IPR environment is also in flux. Our regression analysis shows that legitimate seed from companies associated with CAAS and MDP provides more benefits to farmers than *Bt* seed from unauthorized CAAS or MDP dealers. Such results imply that stronger IPRs could increase benefits to farmers and, at the same time, increase seed companies' profits and innovators' royalties. These changes could also have positive dynamic effects in providing incentives for biotech companies to do more research and develop new technologies.

Finally, seed from seed companies gave better returns than varieties from government agencies, suggesting that it might be time to close down some of the traditional SOE seed operations and promote reform. In other words, more privatization may be beneficial to farmers. It is interesting to note that the *Bt* seed from seed companies had some advantage over farmer-saved seed. But if farmers did not have the cash needed to buy seed, they could still get far superior performance from saved *Bt* seed than from reverting to seed of conventional cotton varieties.

China's experience has shown that small-scale, poor farmers have benefited from GM crops in many ways and can continue to do so. However, setting up and running a sustainable, safe and efficient GM breeding and production programme is not easy. The necessary investments in institutions, R&D and production are immense and require time to develop. While China's progress in almost all dimensions is admirable and holds lessons for other developing countries, it still has a long way to go. Weaknesses and attempts to overcome shortcomings contain many lessons for nations that seek to build a modern agricultural biotechnology system.

Notes

1   Monsanto has patents on several genes that are important in the production of GM plant varieties (henceforth, the Monsanto gene). The Monsanto genes are legitimately managed by two joint venture seed companies set up originally by Monsanto, Delta and Pine Land and a provincial seed company (Jidai in Hebei and Andai in Anhui – henceforth called Jidai for convenience).

2   The two *Bt* varieties are resistant to three stem borers in China: *Tryporyza incertulas* Walker, *Chilo suppressalis* Walker and *Cnaphalocrocis medinalis* Guenee.

3   The hybrid GM II-Youming 86 contains the *CPTi* gene which provides resistance to six pests, the same pests that are targets of varieties containing *Bt* plus *Sesamia inferens* Walker, *Parndra guttata* Bremeret Grey and *Pelopidas mathias* Fabricius.

4   They have produced a draft sequence of the rice genome for the most widely cultivated subspecies in China, *Oryza sativa* L. ssp *indica*, by whole-genome shotgun sequencing.

5   This section is mainly based on Huang et al (2002a, 2002b, 2003a, 2003b, 2004 and 2005a) and Pray et al (2001, 2002). Different papers reflect the results from different years of surveys and crops (e.g. *Bt* cotton and GM rice).

6   Personal communications with MOST officials. The slower progress of stress tolerant varieties should not be surprising or seen as a sign of failure. Stress tolerance has a more complicated mechanism that involves many metabolic pathways between the plant and its environment.

7   In contrast, farmers in regions with higher incomes have tended to produce more vegetable, fruit and other high-value crops.

8   Under the CAS there are at least seven research institutes and four NKLs that focus on agricultural biotechnology. Research institutes within the Chinese Academy of Forestry (CAFo) under the State Forest Bureau and numerous universities (i.e. Beijing University, Fudan University, Nanjing University, Central China Agricultural University, and China Agricultural University) under the Ministry of Education (MOE) are examples of other institutions conducting agricultural biotechnology research. There are seven NKLs located in seven leading universities conducting agricultural biotechnology or agriculturally related basic biotechnology research. Other public biotechnology research efforts on agriculturally related topics include agro-chemical (e.g. fertilizer) research by institutes in the State Petro-Chemical Industrial Bureau.

9   The membership is characteristically created around areas of expertise. For example, 29 of the members are responsible for making decisions and making recommendation on GM plants; nine members are dedicated to examining recombined microorganisms for plants; 12 are responsible for GM animals and recombined microorganisms for animals; and 6 for GM aquatic organisms. All these members work only part-time for BC and are mainly composed of scientists from different disciplines, including agronomy, biotechnology, plant protection, animal science, microbiology, environmental protection and toxicology. A few members also have positions within the MOA and other agricultural-oriented government agencies.

10 The first list of agricultural GM products that were required to be labelled included 17 products from five crops. They are soybean seeds, soybeans, soy flour, soy oil, soy meal; maize seeds, maize, maize oil, maize flour; rape seeds for planting, rape seed, rape seed oil, rape seed meal; cotton seeds for planting; tomato seeds, fresh tomatoes and tomato sauce.

11 In addition, the new regulations state that any *Bt* cotton variety (including

variety line) receiving a production safety certification (or commercializa-tion) from any region can directly apply for a safety certificate in one prov-ince in another cotton-ecological region.

12 One variety, GM Xianyou 63, was created to be resistant to rice stem borer and leaf roller by inserting a Chinese-created *Bt* gene. The other variety, GM II-Youming 86, was also created to be resistant to rice stem borers, but in this case the resistance was created by introducing a modified cowpea tripsin inhibitor (CpTI) gene into rice. The insect-resistant GM varieties entered pre-production trials in 2001.

13 The French acronym for the organization that manages the international treaty that protects plant breeders' rights.

14 Monsanto, Delta and Pine Land joint venture ('foreign seed').

References

BRI (Biotechnology Research Institute) (2000) 'Research achievements of bio-technology', Working Paper, BRI, Chinese Academy of Agricultural Sci-ences, Beijing

Cheng, Z., He, X. and Chen, C. (1997) 'Transgenic wheat plants resistant to barleyyellow dwarf virus obtained by pollen tube pathway-mediated trans-formation', *Chinese Agricultural Science for the Compliments to the 40th Anniver-sary of the Chinese Academy of Agricultural Science*, China Agricultural SciTech Press, Beijing, pp98–108

Fang, X., Cheng, D., Xu, J., Xu, R. and Fan, T. (2001) 'Commercial implemen-tation of intellectual property rights of Chinese transgenic insect resistant cotton with *Bt* gene and *Bt*+CpTI genes,' *Journal of Agricultural Biotechnology*, vol 9 (2), pp3–106

GRAIN (Genetic Resources Action International) (2001) '*Bt* cotton through the back door', *Seedling*, vol 18 (4), December, GRAIN Publications, www.grain.org/publications/seed-01-12-2-en.cfm

Hu, R., Pray, C., Huang, J., Rozelle, S., Fan, C. and Zhang, C. (2005) 'Intellec-tual property, seed and biosafety: Who could benefit from policy reform?', Working Paper, Centre for Chinese Agricultural Policy, Chinese Academy of Sciences, Beijing

Huang, D. (2002) 'Research and development of recombinant microbial agents and biosafety consideration in China', Paper presented at 7th International Symposium on the Biosafety of Genetically Modified Organisms, 10–16 October, Beijing

Huang, J. and Wang, Q. (2003) 'Agricultural biotechnology development and policy in China', *AgBioForum*, vol 5 (3), pp1–15

Huang, J., Qiao, F., Zhang, L. and Rozelle, S. (2000) 'Farm pesticide, rice production, and the environment', EEPSEA Research Report 2001-RR3, IDRC, Singapore

Huang, J., Wang, Q., Zhang, Y. and Falck-Zepeda, J. (2001) 'Agricultural bio-technology development and research capacity in China', Working Paper, Centre for Chinese Agricultural Policy, Chinese Academy of Sciences, Beijing

Huang, J., Rozelle, S., Pray, C. and Wang, Q. (2002a) 'Plant biotechnology in China', *Science*, vol 295, pp674–677

Huang, J., Hu, R., Rozelle, S., Qiao, F. and Pray, C. (2002b) 'Transgenic varieties and productivity of smallholder cotton farmers in China', *Australian Journal of Agricultural and Resource Economics*, vol 46 (3), pp367–387

Huang, J., Rozelle, S. and Pray, C. (2002c) 'Enhancing the crops to feed the poor', *Nature*, vol 418, pp678–684

Huang, J., Hu, R., Fan, C., Pray, C. and Rozelle, S. (2003a) '*Bt* cotton benefits, costs, and impacts in China', *AgBioForum*, vol 5 (3), pp1–14

Huang, J., Hu, R., Pray, C., Qiao, F. and Rozelle, S. (2003b) 'Biotechnology as an alternative to chemical pesticides: A case study of *Bt* cotton in China', *Agricultural Economics*, vol 29, pp55–67

Huang, J., Hu, R. and Rozelle, S. (2003c) *Agricultural Research Investment in China: Challenges and Prospects*, China's Finance and Economy Press, Beijing

Huang, J., Li, N., and Rozelle, S. (2003d) 'Trade reform, household effect, and poverty in rural China', *American Journal of Agricultural Economics*, vol 85 (5), pp1292–1298

Huang, J., Hu, R., van Meijl, H. and van Tongeren, F. (2004) 'Biotechnology boosts to crop productivity in China: Trade and welfare implications', *Journal of Development Economics*, vol 75, pp27–54

Huang, J., Hu, R., Rozelle, S. and Pray, C. (2005a) 'GM rice in farmer fields: Assessing productivity and health effects in China', *Science*, vol 308, pp688–690

Huang, J., Hu, R., Zhang, H., Pray, C. and Falck-Zepeda, J. (2005b) 'GMO biosafety management and regulatory costs: A case study in China', Working Paper, Centre for Chinese Agricultural Policy, Chinese Academy of Sciences, Beijing

James, C. (2004) *Global Status of Commercialized Biotech/GM Crops: 2004*, International Service for the Acquisition of Agri-Biotech Applications, ISAAA Briefs No 32 –2004, available at www.isaaa.org

Keeley, J. (2003) *The Biotech Developmental State? Investigating the Chinese Gene Revolution*, Biotechnology Policy Series No 6, Institute of Development Studies, University of Sussex, Brighton

Li, N. (2000) 'Review on safety administration implementation regulation on agricultural biological genetic engineering in China', paper presented at the China-ASEAN Workshop on Transgenic Plants, 30 July–5 August, Beijing

MOA (Ministry of Agriculture) (1990) 'The guideline for the development of science and technology in middle and long terms: 1990–2000', MOA, Beijing

MOA (1999) 'The application and approval on agricultural biological genetic modified organisms and its products safety', Administrative Office on Agricultural Biological Genetic Engineering, no 4, MAO, Beijing

MOST (Ministry of Science and Technology) (1990) 'Biotechnology development policy', China S&T Press, Beijing

MOST (2000) 'Biotechnology development outline', MOST, Beijing

NCBED (National Centre of Biological Engineering Development) (2000) 'The research progress in biotechnology', *Biological Engineering Progress*, vol 20, special issue

NSBC (National Statistical Bureau of China) (2003) *Statistical Yearbook of China*, China's Statistical Press, Beijing

Paarlberg, R. L. (2001) *Governing the GM Crop Revolution: Policy Choice for Developing Countries*, 2020 Vision Discussion Paper 16, International Food Policy Research Institute, Washington, DC

Peng, Y. (2002) 'Strategic approaches to biosafety studies in China', paper presented at the 7th International Symposium on the Biosafety of Genetically Modified Organisms, Beijing, October 10–16

Pray, C. and Fuglie, K. (2001) 'Private investments in agricultural research and international technology transfer', in *Asia ERS Agricultural Economics Report No. 805*, November, www.ers.usda.gov/publications/aer805/

Pray, C., Ma, D., Huang, J. and Qiao, F. (2001) 'Impact of *Bt* Cotton in China', *World Development*, vol 29, pp813–825

Pray, C., Huang, J. and Rozelle, S. (2002) 'Five years of *Bt* cotton in China: The benefits continue', *The Plant Journal*, vol 31 (4), pp423–430

SDPC (State Development and Planning Commission) (2003) *Development of Biotechnology Industry in China – 2000*, Chemical Industry Press, Beijing

SSTC (State Science and Technology Commission) (1990) *Development Policy of Biotechnology*, The Press of Science and Technology, Beijing

Yu, J., Hu, S., Wang, J., Wong, G. K. S., Li, S., Liu, B., Deng, Y., Dai, L., Dai, Y.,Zhang, Z., Cao, X., Cao, M., Liu, J., Sun, J., Tang, J., Chen, Y., Huang, X., Lin, W., Ye, L. C. and Tong, W. (2002) 'A draft sequence of the rice genome (*Oryza sativa* L. ssp. *Indica*)', *Science*, vol 296 (5565) pp79–108

Zhang, X., Liu, J. and Zhao, Q. (1999) 'Transfer of high lysine-rich gene into maize by microprojectile bombardment and detection of transgenic plants', *Journal of Agricultural Biotechnology*, vol 7 (4), pp363–367

9

India: Confronting the Challenge – The Potential of Genetically Modified Crops for the Poor

Bharat Ramaswami and Carl E. Pray[1]

Introduction

Can GM crops reduce poverty? Is this a likely outcome? This paper examines evidence of the impact of GM crops in India and what light, if any, it throws on these questions. Indian experience began in 2002 when the first GM crop was approved for commercial release, namely three hybrid varieties of *Bt* cotton. In 2004 and 2005 the government granted permission for the release of several other hybrid varieties of *Bt* cotton and more approvals are expected for the 2006 season. In addition, an unauthorized *Bt* cotton variety discovered in farmers' fields at the end of 2001 continues to be used, particularly in the states of Gujarat, Punjab and Andhra Pradesh. No other GM crop has been commercially released. Table 9.1 shows trends in areas planted to 2004/2005.

For economists, impacts that matter most are those affecting the economic welfare of growers, consumers and seed market agents such as suppliers. However, these are in some sense 'reduced form' impacts – the outcome of various processes including basic research, technology adaptation, biosafety regulatory procedures and their enforcement, seed pricing and competition in the seed market. As the government naturally has a large presence in these activities, its policies and the institutional mechanisms devised to formulate and implement them are among the 'structural' factors that explain the reduced form impacts, though government policies have been vigorously contested. While this chapter does not offer a 'deep' explanation, it attempts to demarcate the constituencies that have pressured policies and their enforcement.

Table 9.1 *Area planted with Bt cotton in India (acres)*

	2000/2001	2001/2002	2002/2003	2003/2004	2004/2005
NB 151 F$_1$ and F$_2$	200	6000	100,000	600,000	2,000,000
MMB	–	–	100,000	200,000	800,000
Rasi					200,000
Total *Bt* cotton			200,000	800,000	3,000,000

Note: 1 acre = 0.405 hectares.
Source: Pray et al (2005)

Poverty reduction and GM crops: The links

The 2004 report of the Nuffield Council on Bioethics drew an analogy with the Green Revolution to delineate how GM crops could reduce poverty (Nuffield Council on Bioethics, 2004). Across much of Asia and Latin America available farmland is largely exhausted while expansion of non-farm employment opportunities in industry requires large investments in equipment, buildings and infrastructure. Thus, higher productivity and greater employment in agriculture is the most effective route to poverty reduction. The Green Revolution created employment for landless agricultural workers, increased yields for small farmers and reduced prices of food staples for poor consumers (Lipton and Longhurst, 1989). Now, as conventional plant breeding possibilities near exhaustion, use of GM crops could improve yields of food staples and other crops grown by the poor.

In India, the proportion of rural population living in poverty declined from above 50 per cent in the mid-1970s to about 31 per cent by the end of the 20th century.[2] During most of this period, the non-farm sector grew at twice the rate (or more) of the agricultural sector. Nevertheless, rural economic growth was found to have significant impact on reducing urban and rural poverty, while urban growth affected rural poverty very little (Ravallion and Datt, 1996). Higher farm yields is the key variable that reduces rural poverty and increases wage earnings (Datt and Ravallion, 1998).

Not surprisingly, poverty is highly correlated (inversely) with the level of agricultural earnings (Kijima and Lanjouw, 2005). As agricultural wages tend to reflect earnings of workers in other sectors, changes in agricultural wages are good indicators of changes in poverty. In India, real daily agricultural earnings increased by 69 per cent between 1983 and 1999 (Eswaran et al, 2006). In a simple two-sector equilibrium model, agricultural wages are determined by total factor productivity in agriculture and in the non-farm sector (Eswaran and Kotwal, 1993). So what have been the relative contributions of these two factors in explaining earnings increase? Despite its faster growth, the contribution of increased non-farm productivity was found to be quite limited (Kijima and Lanjouw, 2005; Eswaran et al, 2006). It was concluded that while expansion of non-farm employment does put some pressure on the agricultural labour market and help to raise agricultural wages, its impact on poverty reduction is minimal.

The Nuffield Council report emphasized the relevance of GM crops where non-farm sector growth is expensive and difficult to achieve. However, agricultural productivity growth can be central to reducing poverty even when non-farm growth is rapid. While the non-farm sector might become more important in the future, it seems very unlikely that it will be able to absorb the large numbers of poorly educated members of the labour force currently employed in agriculture.

The pro-poor potential of GM crops is more often than not asserted through Malthusian arguments that increased population pressure requires more productive technologies (Herring, 2005). However, it is well known that hunger is equally an outcome of unequal entitlements to food. The pro-poor potential of GM crops is more properly seen in improving agricultural productivity and rural incomes. Even if population growth rates are low, agricultural productivity growth can be critical to poverty reduction (Eswaran and Kotwal, 1993).

Government policy: Objectives, priorities, commitment

Biotechnology has received explicit and special attention in Indian public policy. In 1986 the government set up the Department of Biotechnology (DBT) in the Ministry of Science and Technology, giving this field the same status as atomic energy and space exploration within its science portfolio. The DBT has invested resources in education, training, research labs and networks and in its official documents it lauds biotechnology for its potential in agriculture, healthcare and other areas. It is seen as a sector where India could possess comparative advantage and be competitive globally (GoI, 2005).

The potential of crop biotechnology is seen with reference to limited natural resources, especially land, low productivity in dryland farming areas (bypassed by the Green Revolution), and loss of momentum in yield advances (Sharma et al, 2003; GoI, 2005).[3] As the official in charge of India's agricultural research programme recently asserted, 'the search, characterization, isolation and utilization of new genes through application of biotechnology are essential for the revitalization of Indian agriculture' (Rai, 2006).

Nevertheless, official support in practice has been sporadic and modest. In 2004, the government accepted a strategy for agricultural biotechnology that has two essential components (GoI, 2004). The first defines the scope of crop biotechnology by listing applications to be discouraged. GM research is not to be undertaken on exportable crops. Transgenes will not be commercialized in certain parts of the country defined as 'agro-biodiversity sanctuaries' or 'organic farming zones'. Low priority is to be given to biotechnology applications that are potentially labour-saving (such as herbicide tolerant traits).

The second component sets priorities, calling for high priority to be accorded to biotech applications that do not involve GM such as biopesticides, biofertilizers, bio-remediation agents, plant tissue culture and molecular assisted breeding. It also lists the traits and crops that deserve priority GM research. The strategy's priorities largely overlap those of Grover and Pental's 2003 survey of the research priorities of agricultural scientists involved in improvement of 12 major field crops, suggesting a consensus among the research community. Breeding for resistance to biotic stresses, pests and pathogens are major objectives for all crops. While improving water use efficiency and GM approaches to abiotic stresses are also recognized as deserving high priority, payoffs here are seen as less immediate.

For each specific crop, GM approaches are suggested for problems that are intractable using conventional breeding techniques. For instance, in the case of rice, conventional plant breeding is regarded as adequate for providing resistance to blast, bacterial leaf blight, tungro virus, gall midge, brown plant hopper and whiteback plant hopper. However, germplasm resources for stem borer, leaf folder, sheath blight and sheath rot are deemed as inadequate and requiring GM techniques. Between the two major cereal crops, rice receives higher priority for GM approaches as most of the biotic stresses in wheat can be dealt with by conventional breeding technologies.

Public sector research: Agenda and results

Situated outside the public sector agricultural research institutes, the DBT funds plant biotechnology projects both within and outside of these institutes. It also occupies a central position in the regulatory apparatus (discussed below). Thus, a wider range of expertise than could be found in traditional centres of plant breeding has been applied to plant biotechnology. This is a positive development in that it has broken the long-standing institutional monopoly of the public sector in agricultural research. However, it also gives rise to new concerns. Because of their distance from the final users of new biotechnology (i.e. the farmers) those engaged in public sector agricultural research must constantly redefine their priorities and allocate resources accordingly, especially the public sector researchers outside the specialized agricultural research institutes. The researchers within these facilities have the advantage of links with allied plant disciplines (including traditional plant breeding) and agricultural extension services, at least in principle.

The DBT supports research projects at different research institutes and agricultural universities throughout the country. It has also established specialized centres for plant biotechnology research. Specific activities funded include basic research in plant molecular biology and genomics, particularly rice genomics, in collaboration with the international genome sequencing programme. Other 'knowledge-building' types of work include tagging of quality traits in rice, wheat and mustard, and molecular methods for heterosis breeding.

In 2003, 47 projects in the public sector aimed at developing transgenes in various crops, 33 of them with resistance to insects, viruses or fungal infections. Among these, 14 projects aimed at using a *Bt* gene to develop insect resistant varieties of cotton, potato, tobacco, rice and vegetables. Other projects aimed at transgenes with male sterile and restorer lines for hybrid seed production, to delay fruit ripening, to enhance nutrition, to withstand moisture stress or flooding, and to supply edible vaccines. About half of the projects involved rice or vegetables. Other crops researched included chickpea, mustard/rapeseed, tobacco, cotton and blackgram. Because of complex genetic mechanisms, field deployment of abiotic, stress-tolerant GM crops is still regarded as distant (Grover et al, 2003). Here, more funding for research as well as collaboration among plant molecular biologists, crop physiologists and agronomists would be required.

India's public sector research programme has been criticized for spreading resources too thinly and not orchestrating a concerted research effort with select crops and well-defined goals. In 2002/2003, the annual DBT budget for crop biotechnology was only about US$3 million and total spending planned for five years (starting in 2002) was no more than US$15 million (Sharma et al, 2003). Not a single product from the public research system is in large-scale trials or close to commercialization.

Several factors seem to be responsible. First, within the traditional agricultural research institutions expertise in plant biotechnology has remained limited (Pental, 2005) and there has not been an aggressive move to acquire it. Second, the development of transgenes for commercial use requires teams proficient in various disciplines such as agronomy, plant breeding, plant pathology, entomology and biotechnology. The public sector has failed to develop such coordinated approaches. Third, the public sector has not incorporated regulatory know-how in the design of its research projects (Pray et al, 2005). Research budgets do not earmark funds for regulatory costs and delays in the regulatory process are common. A case in point is the work on insect resistance for *basmati* rice, an exportable with major markets in Europe and the Middle East. A regulatory advisor could easily have anticipated the project's difficulties in this area.

Biotechnology in the private sector

Private sector investments in biotechnology have been largely in cotton, rice and vegetables, and in a single trait – insect resistance – through *Bt* genes.

The exception was Bayer's research on genetically modified hybrid mustard. Interviews with a large number of seed/biotech firms in 2003 and 2004 (Pray et al, 2005) found that the regulatory climate had induced private firms to shift research and technology transfer priorities away from rice, vegetables and mustard toward cotton.

Both global and local factors caused the decline in rice biotech research. Globally, multinational biotech firms have reduced their research on GM rice and in India, a centre for rice biodiversity, there are special ecological concerns about it. If a GM rice variety is exportable, or if it cannot be segregated from exportable varieties, regulators have to take this into account. Bayer withdrew from commercialization of GM mustard in 2003 because of continued regulatory costs and uncertainty about whether this product would ever be approved. Among vegetables, Mahyco's *Bt* eggplant, in large-scale trials, could be the first food crop to be seriously considered by the regulatory system.

Cotton biotech research, by contrast, is on the rise. No major company has dropped out and new companies are starting applied and basic biotech programmes. In India, the first approvals to *Bt* cotton were given to three hybrids released by Mahyco Monsanto Biotech (MMB), a joint venture between an Indian seed company, Mahyco, and the major American biotech, Monsanto. These hybrids contained the *Bt* gene *Cry1Ac* owned by Monsanto under the brand name Bollgard. Subsequently, MMB sub-licensed the gene to other firms in India (20 as of April 2005) allowing them to incorporate it into their cotton hybrids. Monsanto is pushing the next generation of *Bt* technology – Bollgard II – which stacks *Cry1Ac* and *Cry2Ab*, through the Indian regulatory system.

Non-Monsanto *Bt* genes are still going through the regulatory process. Syngenta has been working with their VIP gene for insect resistance. JK Seeds is using a modified *Cry1Ac* gene developed in collaboration with the Indian Institute of Technology, Kharagpur. Nath Seeds has sourced a *Bt* gene from the Chinese Academy of Agricultural Sciences.

Biosafety regulation: How has it worked?

Indian regulatory institutions have three layers. At the bottom, an institutional biosafety committee (IBC) must be established in any institute using DNA in its research. These committees comprise institute scientists and also a member from the DBT. The IBC can approve research done at the institute unless it involves a particularly hazardous gene or technique. That type of research must be approved by the Review Committee on Genetic Manipulation (RCGM), the next layer of the system.

The RCGM, within the DBT, regulates agricultural biotech research up to large-scale field trials. It requests food biosafety, environmental impact and agronomic data from applicants wishing to do research or conduct field trials and gives permits to import GM material for research. It consists primarily of scientists, including agricultural scientists, and can request specialists to review cases. Its Monitoring-cum-Evaluation Committee monitors field trials of GM crops.

The Genetic Engineering Approval Committee (GEAC), under the Ministry of Environment and Forests, is the agency that gives permits for commercial production, large-scale field trials and imports of GM products. Although scientists are members of this committee, bureaucrats representing different ministries predominate.

Experience with regulation is exemplified by the first product that was commercialized. It contained the first event to be approved, the *Bt* gene, *Cry1Ac* from Monsanto, which was inserted in three cotton hybrid cultivars (MECH 12, MECH 162 and MECH 184) belonging to the Indian seed company, Mahyco. The first biosafety tests were done in 1997, after backcrossing, and approval for commercial release came five years later when the varieties were accepted for cultivation in southern, western and central India for a three-year period.

As the first GM product to go through the regulatory system, MMB *Bt* cotton attracted media attention. Several Indian and international NGOs opposed the application and the regulatory process was repeatedly challenged. On the basis of environmental and biosafety tests and field trials, MMB sought commercial release in 2001. However, the regulator rejected its request and asked MMB to conduct field trials at 40 locations under the direct supervision of the public sector research body, the Indian Council of Agricultural Research (ICAR). This cautious stand of the regulator and its involvement of ICAR seemed aimed at deflecting the pressure from NGOs, suspicious of the data generated from Mahyco's experiments. According to newspaper reports, the scientist members of GEAC, favouring approval, were outvoted by the bureaucrats (Jain, 2001).

This controversy led to the regulator requiring at least a year of ICAR-supervised field trials for all subsequent product approvals. Varietal testing therefore goes through small-scale trials with RCGM and large-scale trials with both GEAC and ICAR. This has indeed diffused challenges to regulatory decisions, but it has also highlighted the role of large-scale field trials in the regulatory process. The primary purpose of those conducted by ICAR is not environmental but agronomic and economic. It is assumed that farmers are unable to compare alternative varieties and must therefore be protected from potentially disastrous choices. Thus, the regulator is not merely a guarantor of the food and environmental safety of GM products but also of the agronomic and economic performance of GM crops. The redefinition of the job testifies to the pressures exerted by GM crop opponents.[4]

The 'illegal' seeds

Regulators have also had to cope with pressures from farmers. In November 2001 they discovered that some farmers in Gujarat had planted a cotton hybrid containing the *Cry1Ac* gene. This was NB 151, a variety registered with the Gujarat government as a conventional hybrid, but actually illegal as it had not been approved for release by the biosafety regulators. Multiplication and distribution of illegal seed occurs through an underground network of seed producers, small seed companies and their agents. Despite government prosecution of the guilty firm and its officials, plantings of illegal *Bt* cotton have spread across Gujarat and to other parts of India, notably Punjab.

While the state government is responsible for prosecuting violations of biosafety law, in the face of strong farmer support for illegal seeds, it has chosen to turn a blind eye. Seed law exempts farmer-to-farmer exchange of seed from inspection and this has allowed the state government to claim ignorance of the extent of illegal plantings. Moreover, illegal seed sellers try to mask their sales as seed exchange; illegal seeds are often sold loose in packets without a company seal and with no bill of sale.

The discovery of the illegal plantings with the complicity of the state government in late 2001 probably reassured GEAC that it was correct to approve the MMB hybrids in 2002. The GEAC also faces direct pressures from farmer representatives, including chief ministers of agriculturally prosperous states like Punjab (Jain, 2002). The initial approvals of the MMB varieties did not extend to Punjab and, worried by the illegal plantings, state government officials pressed the regulators for approval of varieties for their region. The latter appear to have responded, wishing to combat the spread of illegal seed. Since 2004 they have approved several other *Bt* hybrids, some from MMB, but most from other seed companies who have licensed the *Cry1Ac* gene from MMB. The regulators have used this fact to do away with food safety and environmental tests, basing their approval on large-scale field trials for agronomic and economic performance. Approval of a cotton hybrid with a *Bt* gene other than *Cry1A* is expected in 2006.

Implementation process: Political economy dynamics

The normative view of biotech regulation is that it is a process of risk assessment based on rigorous science. However, as the Indian experience attests, it is an intensely political process, contested at many levels. NGOs and civil society organizations have debated and questioned the direction of agricultural technology and forms of corporate control. Farmers have challenged the enforcement of biosafety laws that they consider out of touch with their interests. Corporations use their public relations officials to influence the process. Three government departments – biotechnology, environment and agriculture – are actively involved, each with its own interests.[5] The regulatory process has had to deal with turf disputes between scientists with different types of expertise (for example, biotech lab experience, agricultural field experience) as well as between scientists and bureaucrats.

With so many pressures, the regulatory process is subject to delays and not entirely predictable. Compliance costs of four products that went through the regulatory system or are still under regulation have been surveyed (Pray et al, 2005). These were MMB's first *Bt* cotton hybrids, Bayer's GM mustard hybrid and *Bt* eggplant and high-protein potato from public sector research institutes. Compliance costs were found to be high for MMB and Bayer. In the case of MMB, pre-approval costs were about US$1.8 million, of which US$300,000 was spent on field trials. (The largest value of cotton seed sales from any single firm is approximately US$30 million per year.)

Bayer's compliance costs were even higher, in the range of US$4–5 million. The genes used to produce hybrid mustard have been used in canola to produce hybrid canola cultivars in Canada and the US, where they have cleared the biosafety regulations. However, use of these genes in mustard has not been commercialized anywhere in the world. Because of continued costs, uncertainty about whether GM mustard would ever be approved and the market potential for this product, Bayer decided not to continue trying to commercialize it in India.

By contrast, compliance costs have not been a major constraint to research or commercialization efforts in the public sector. Regulatory delays have been the principal issue. In the case of *Bt* eggplant from the Indian Agricultural Research Institute, small-scale multi-location trials were delayed by three years. When a project has the full support of the DBT, the time and cost of regulation can be reduced. For example, regulatory costs have been minimal for high protein potato research at the Centre for Plant Genomics Research in New Delhi, often cited by the previous head of the DBT as exemplifying the consumer benefits from GM technology, though the product has not yet been approved for commercial release.

Private companies have also been polled about the costs of meeting biosafety regulations (Pray et al, 2005). These vary widely based on the type of crop; whether the gene already had been approved by regulators in India or elsewhere (i.e. in the US or Europe); and whether the tests could all be completed in India. Other differences may occur if companies wish to do more research than is required by Indian regulators in order to document certain qualities of the crops other than those required in the country.

The least expensive new events (costing about US$100,000) will be in non-food crops like cotton, involving events that have been in commercial use elsewhere like Monsanto's *Cry1Ac Bt* gene. Much basic information and the results of many field and toxicity/allergenicity tests are available from the US and other countries. US and European companies now spend from US$5–10 million for each new gene, assembling a package of information for regulators and customers in each new country in which they introduce the gene. They then perform whatever additional tests are required, taking into account differences in the way the crop is consumed, local nutritional issues and specific agricultural and environmental conditions. Ethical or political values may also enter the picture (for example, India's requirement that new varieties be tested for 'Terminator' genes, which it prohibits). New events in food crops are likely to cost the most – in the range of US$4 million.

Why are private companies' estimates so high relative to public sector costs? In fact, except for salaries, there is no reason for the latter to be any lower. While there may be an incentive for private sector participants to exaggerate costs in order to lobby for lower ones, public sector costs seem to be substantially underestimated due to the way programmes are managed. For example, public sector research programmes do not budget separately for compliance, salary costs are paid from a general budget, and biosafety tests are often done by other public sector research institutes that charge public sector scientists nominal amounts. It appears that costs reported by the private sector are more accurate.

This gap between private and public sector compliance costs is expected to narrow. As the public sector moves more products through the biosafety process, it seems unlikely that it will remain insulated from regulatory costs. Government labs will begin charging commercial rates for public sector biosafety testing. As internationally certified domestic testing facilities become available, the private sector will be able to avoid the expense of doing tests abroad. The most detailed data available – for Mahyco's *Bt* cotton and Bayer's hybrid mustard – are for the first products that went through the system, with a lot of 'learning by doing'. Future products may not cost as much (although they could cost more if regulation becomes more stringent and more tests are required).[6]

The surplus from *Bt* cotton: distribution of gains among farmers, consumers and seed companies

The area throughout India under cotton fluctuates. It is usually around 22 million acres (9 million hectares) of which the share of hybrids (public as well as private) is estimated to be more than two-thirds (Murugkar et al, 2006). Table 9.1 displays the area under *Bt* cotton hybrids since 2000. It is estimated that *Bt* cotton was being grown on 3 million acres (1.2 million hectares) by the end of 2004/2005. Of this, illegal *Bt* varieties occupied nearly 2 million acres (0.8 million hectares). It is believed that the area under *Bt* cotton in the 2005/2006 season is in the range of 8–9 million acres (3.2–3.6 million hectares), of which illegal seed is thought to account for 5–6 million acres (2–2.4 million hectares). As the cotton growing areas are specialized into hybrid regions and variety regions, and since that division has been slowly changing, it is reasonable to assume that *Bt* cotton hybrids have replaced other hybrids.

The impact of MMB *Bt* cotton on crop yield and farmers' income has been hotly contested. Many farm 'surveys' have been carried out – by the media, NGOs and industry. While these are difficult to evaluate, papers published in academic and policy journals compare the performance of *Bt* cotton with a check variety in terms of yields and use of inputs, estimating the change in farm income due to *Bt* cotton for given output and input costs. Although the procedure seems deceptively simple, there are pitfalls in its execution. First, a survey must be designed so that the selection of growers is truly random and not biased towards a region or other grower characteristic. Second, the correct counter-factual must be identified. In the absence of *Bt* cotton, what would the adopters do? Would they be growing the check variety? Third, a comparison of adopters and non-adopters must control for differences in observable and unobservable characteristics. The easiest way to do this would be to compare *Bt* and non-*Bt* plots of the same farmer. This has been done in a few studies and it would work wherever there are large numbers of partial adopters.

Table 9.2 compares the difference between group means of *Bt* adopters and non-*Bt* adopters across five different studies. Of these, Bennett et al (2004) and Sahai and Rehman (2004) have results for the years 2002 and 2003, giving us comparisons from seven surveys across the years 2001 to 2003. The surveys differ in terms of sample size, states surveyed and whether they control for individual grower characteristics. Among them, Sahai and Rehman's 2004 study stands out as the only one showing a worse performance for *Bt* cotton compared to other commonly grown hybrids. Otherwise, all the papers present a common picture despite differences in methodologies. Net returns to the grower (relative to the non-*Bt* alternative) range from Rs3400 to Rs8800 (US$76–196) per acre. The increase in percentage terms varies from 49 per cent (Bennett et al (2004) for the year 2002) to 480 per cent (Qaim (2003) for the year 2001). The Qaim study uses data from MMB field trials in 2001. The Bambawale et al (2004) analysis uses an experimental setting to compare *Bt* cotton hybrids with non-*Bt* cotton hybrids under similar production practices. All the other studies use data from farmers growing *Bt* cotton under normal field conditions. It is difficult to explain the poor performance of *Bt* cotton in the Sahai and Rehman analysis and how it can be reconciled with the rapid adoption of *Bt* cotton overall. It has been suggested (Naik et al, 2005) that performance of *Bt* cotton has not been uniform across states and that its advantage over non-*Bt* cotton has been minimal in Andhra Pradesh – the state from which Sahai and Rehman draw their analysis.

Taking a conservative view of the performance of *Bt* cotton, let us suppose the return from it relative to non-*Bt* alternatives is Rs2161 (US$48) per acre, the lowest figure in Table 9.2 (except for Sahai and Rehman's). We can interpret it as the average all India figure. From the Bennet et al and the Naik studies, we see that the cost of *Bt* seed for 1 acre is Rs550 (US$12) and that of non-*Bt* seed is Rs500 (US$11). The net surplus to the seed industry from *Bt* cotton is therefore Rs1050 (US$23) per acre. The total surplus per acre generated by *Bt* cotton is the sum of grower returns and seed industry profits, which works out to Rs3211 (US$71) per acre. The share of the seed industry is 33 per cent and the remaining 67 per cent remains with the grower. Table 9.2 suggests that 67 per cent is lower, bound to the share of the grower in the surplus. In terms of aggregate gains, applying the gains to growers to the 2004 diffusion level (1 million acres of legal *Bt*) means an increase in aggregate gains of over Rs2 billion (US$44.5 million). As a proportion of overall farmer income from hybrid cotton, the gains amount to 7 per cent. The above calculations assume

Table 9.2 *Differences between Bt and non-Bt variety*

	Qaim (2003)	Bamba-wale et al (2004)	Bennett et al (2004)	Naik et al (2005)	Sahai and Rehman (2004)	Bennett et al (2004)	Sahai and Rehman (2004)
Year	2001	2002	2002	2002	2002	2003	2003
Sample size (no of growers)	157	NA	2709	341	136	787	136
States	Maharashtra, Madhya Pradesh, Tamil Nadu	Maharashtra	Maharashtra	Maharashtra, Karnataka, Andhra Pradesh, Tamil Nadu	Andhra Pradesh	Maharashtra	Andhra Pradesh
Controls	Yes	Yes	Yes	No	No	Yes	No
Seed + pesticide cost	651	839.68	301.21	213	–	46.15	–
Total cost	1159	940.89	–	1217	983	–	950
Yield (kg)	283	214.98	275.3	168	–70	352.23	0
Revenue	5573	4948.99	5474.49	3378	–2425	8809.72	0
Returns	4414	4010.53	5178.54	2161	–3408	8755.06	–950

Note: all in Rs/acre (except yield, which is kg/acre).

that the additional supply due to *Bt* cotton does not affect prices. As *Bt* cotton diffuses, it will reduce cotton prices. Consumers will benefit and producer gains will therefore not be as much as when prices remained unchanged. However, the sum of consumer and producer benefits will continue to add up to 67 per cent. The exact division of gains between these two groups of agents depends on the elasticity of demand for cotton.

GM cotton seeds market: Is it competitive?

India's cotton seed market consists of three segments: varieties, public bred hybrids and private bred hybrids. By value, private bred hybrids dominate, accounting for 86 per cent of the value of the market. A *Bt* cotton hybrid seed is priced three to four times higher than a non-GM hybrid seed. Therefore, as *Bt* cotton diffuses, the value of the cotton seed market rises rapidly. It is estimated that more than half of the increase in the value of the seed market between 2002/2003 and 2004/2005 was due to *Bt* cotton and projections are that *Bt* seeds will increase the seed market by 22 per cent in 2005/2006 (Murugkar et al, 2006). If most of this increase in value accrues to owners of the technology, would that not become a force for consolidation?

In fact, the rapid growth of the private bred hybrid segment has not been accompanied by greater consolidation. With market growth, more players have come in, eating away at the share of the market leaders. Murugkar et al (2006) show that there are at least 15 firms with successful hybrid products. They argue that when judged by commonly used concentration indices – the entry of new brands, the fluctuation in market leaders and the number of established brands – the hybrid seed market has become more competitive over the last decade.

With *Bt* cotton, the seed industry encompasses a seed market as well as a technology market. As of now, the technology market consists of only one supplier – MMB, which has licensed its *Bt* gene to almost all of the leading cotton seed companies. For a seed company, developing a *Bt* product means a substantial hike in R&D investment. However, that has not constituted an entry barrier as more than 20 firms have licensed *Bt* genes from MMB. Also, not all of these firms yet have products in the market. For instance, in the 2005 season, besides MMB, hybrids from three other firms – Ankurt, Rasi and Nuziveedu were available to growers. Hybrids from other firms are still in large-scale trials awaiting GEAC's approval or at more preliminary stages of testing. Some licensees concluded their agreement with MMB in 2005 and are just beginning to do backcrossing. By contrast, Rasi's agreement with MMB dates from 1998. It conducted large-scale trials in 2002 and 2003 and obtained GEAC's permission to commercialize in 2004. The fact that not all firms started their *Bt* programmes at the same time means that those that got a head start temporarily enjoy monopoly power. GEAC's insistence on agronomic testing (through large-scale trials) favours firms that have already received commercialization approvals. Such testing is not mandatory for non-*Bt* hybrids. Although private firms can get their hybrids 'notified' by having them tested in public sector research trials they have preferred to rely on their own quality systems to build brands and push sales.

While the entry of additional *Bt* hybrids would offer growers more choices, the impact on price would be muted because, by the licensing agreement, all firms pay a fixed sum per packet of seed as trait value to MMB. *Bt* hybrids with non-Monsanto genes are expected to be approved for commercial release in 2006 or 2007. The competition from alternative genes could lead to a more serious impact on the seed price than the competition between hybrids with the MMB gene if the alternative gene providers target a trait value lower than that fixed by MMB. Whether this will happen and to what extent will depend on: first, the performance of these alternatives as compared to MMB's genes, especially Bollgard II, which promises protection against lepidopetran and the rapidly emerging spodoptera pests; and second, MMB's first mover advantage in sub-licensing the Monsanto genes to firms that have some of the best performing hybrids. Even if the alternative gene constructs prove successful, they may not be able to combine with quality germplasm. Thus, the market for the new genes may well be limited by the contractual restrictions of the major seed firms with MMB.

MMB's position as the sole gene supplier is not protected by intellectual property laws. Although India now provides for plant breeders' rights, these have not been operationalized. Even if they are, the private seed industry will be unlikely to utilize them because these rights provide few incentives for innovation (Srinivasan, 2004). As for patent laws, India's compliance with TRIPs norms means that technology suppliers can patent genes. However, the patents office has not yet granted any claims.

MMB has derived a measure of protection for its gene through biosafety laws. As biosafety approvals are obtained for the composite of the gene and the germplasm, hybrids that incorporate MMB's gene but do not go through the biosafety process are illegal. While this has not stopped the diffusion of illegal *Bt* seeds, it has led the seed companies wishing to work within the law (all of the established firms with branded products) to either deal with MMB or consider an alternative *Bt* strategy. At this point, most of the firms have chosen to license the *Bt* technology from MMB.

MMB would have gained even more from its legal monopoly but for the illegal *Bt* varieties that originated and are still dominate in Gujarat and have also spilled over into Maharashtra, Punjab and Andhra Pradesh. In the 2004 season, illegal *Bt* was priced anywhere between Rs800 (US$18) and Rs1200 (US$27) compared to Rs1600 (US$36) for a packet of legal seeds.[7] With its seemingly effective performance and its lower price, illegal *Bt* is a threat to legal seed, *Bt* or otherwise. This threat is particularly acute for non-*Bt* hybrids. With legal *Bt*, the non-*Bt* market has some protection because of the large difference in seed price. With illegal *Bt*, there is much less protection. In Gujarat, for instance, the market leader, Vikram Seeds, lost its non-*Bt* market rapidly because of illegal *Bt*.

A huge concern for the suppliers of legal *Bt* seeds is whether the illegal seeds will wipe out their market. The geographical spread of illegal seeds could be limited by its underground nature as illegal *Bt* seeds are also hybrid seeds, not reproduced by farmers but produced and distributed by a network of seed producers and distributors. The production of hybrid cotton seed requires skill, experience and access to parent lines. Gujarat has a long history of cotton seed production and some seed producers have a male parent with a *Bt* gene. NB 151 is now a generic name for illegal *Bt*. It is believed that the male parent (with the *Bt* gene) used in this variety has been crossed with a variety of female lines to generate many different versions of illegal *Bt*, often well adapted to local environments. For illegal *Bt* to diffuse widely, either seed production has to migrate or the seeds themselves have to be distributed. The second possibility is easier to imagine but even here, seed suppliers cannot use normal commercial channels to deal with first time buyers and transactions cannot be made with banking facilities but must be based on trust. An additional difficulty is that the seed cannot be branded and illegal seed producers therefore have no formal means of communicating quality to growers outside their traditional areas of operation.

Revisiting the impact of GM crops on the poor

The hope is that GM crops will revitalize crop productivity, increase the incomes of small farmers and landless workers and reduce poverty. How realistic is this possibility? And what will be needed to make it happen? In the view of crop improvement experts, GM technologies are the only way to deal with many kinds of biotic stresses in numbers of crops (Grover and Pental, 2003). Using them could reduce crop losses and significantly increase productivity, especially in dryland agriculture. GM solutions to abiotic stresses (for example, moisture stress, salinity) would have major impacts. But these require more basic research and a longer timeline is forecast for their development.

As mentioned, India's private sector has a strong presence in the distribution and marketing of seeds as well as in the development of new varieties of certain crops. The diffusion of *Bt* cotton has been the handiwork of private agents, with legal *Bt* backed by large firms with technical and marketing prowess. Unofficial *Bt* has spread on the strength of a network of skilled seed producers, small companies and their agents. The demand for both kinds of seeds has been strong because of their considerable advantages over conventional hybrids in protecting yields from pest losses. Thus, if farmers perceive gains from using certain types of seeds, the private sector has sufficient capabilities to supply them. Constraints to the adoption of beneficial GM crops do not arise from distribution.

What is of concern is appropriability. Private sector activity is confined to hybrid seed. Although India has plant breeders' rights, it is unlikely to stimulate any private sector interest in open pollinated varieties because the rights protection does not apply to seed saved or exchanged by farmers. For poverty impacts, crop productivity must rise in the major food staples, namely, the open pollinated crops of rice and wheat as well as the essentially self-pollinated grain legumes (chickpeas, pigeon peas, mung beans, groundnut, soybeans) that are extensively grown in the rainfed and dryland areas. Except in the case of wheat, scientists believe that GM technologies are essential to develop varieties resistant to pests and pathogens (Grover and Pental, 2003). Hence, governments need to solve the appropriability problem. Conventionally, what has been done is to invest the responsibility of public goods type research with the public sector, as was done with the Green Revolution. In India, however, the public sector is not yet well equipped to play this role with regard to GM crops.

Several difficulties will have to be overcome. First, the level of funding is presently too low (especially in relation to potential benefits) to support initiatives on a large scale. Second, funds need to be deployed in a focused and sustained manner. Third, there is a lack of relevant expertise within the public agricultural research sector. Public–private partnerships have been proposed in this context though none yet exist. Fourth, most of the public sector is not yet in tune with regulatory demands.[8]

If the private sector will not invest in R&D for a large number of crops, and if it is unable to take up the slack, what can be done? Experts distinguish between 'push' and 'pull' programmes to encourage R&D (Kremer and Zwane, 2005). Public sector research has typically been of the push kind. While push programmes are appropriate for basic research it is argued that they do not work as well in inducing development of products that receive wide adoption among farmers. Kremer and Zwane advocate pull programmes where the reward to technology owners is tied to adoption. Clearly, this is an attractive option whenever it is infeasible to create intellectual property rights, as with seeds.

As regards regulation, India is on a learning curve. While the regulatory process for initial products was costly and suffered delays, a more streamlined one should apply henceforth. However, this process will continue to reflect pressures from both anti- and pro-GM voices, resulting in uncertainty. If regulatory costs remain high, the private sector will focus on hybrids with large markets. Large firms that can manage regulatory procedures and expenses are unlikely to fear them. The first entrant into the *Bt* cotton market in India earned non-competitive profits but the growers were still better off.

A tricky issue is how the government should deal with illegal *Bt* seeds. By many accounts, these seeds have done well in Gujarat where they are well adapted to local growing conditions and have been backed by an effective if informal governance system reassuring farmers of quality (Herring, 2005; Lalitha et al, 2006; Murugkar et al, 2006). Not surprisingly, NB 151 and its variants have been widely adopted and a government that tried to enforce the law would suffer politically. It cannot be economically efficient to deprive farmers of a well-adapted variety. By contrast, illegal seeds have reduced the private returns to MMB's R&D. Illegal GM seeds are not unique to India; they are rampant in Brazil and China as well. Even with IPRs and biosafety laws, weak enforcement in developing countries will reduce the ability of private innovators to appropriate the gains, which in turn affects the incentives of biotech firms to develop products for these countries. The arguments of Kremer and Zwane (2005) suggest that governments could rectify this somewhat through pull programmes of research. This implies that the Indian government should stop worrying about the diffusion of illegal seeds (which are as safe and as proven in farmers' fields as the legal varieties) and compensate MMB in relation to the social gains from such diffusion.

The release of the *Bt* technology in India was accompanied by refuge policies whereby farmers were required to plant non-*Bt* cotton around *Bt* cotton. The need for such policies has been questioned where mixed cropping could provide alternative hosts for pests. However, assuming some kind of refuge restriction is desirable to manage resistance to the *Bt* toxin, how could compliance be ensured when it is not in the interest of the individual grower? This is an issue in any society; even in the US, compliance has been imperfect (Buttel et al, 2005). The problem cannot be easier in India where potential offenders are poor and numerous and enforcement capacity is weak. Finally, externalities in agriculture are not due to GM technologies alone. They frequently arise in many other contexts, including groundwater depletion and pest management. In the din of GM politics, however, such issues have been pushed to the background.

Notes

1   For valuable comments, we thank Sakiko Fukuda-Parr and participants at the seminar, Making GM Crops Work for Human Development: Socio-Economic Issues and Institutional Challenges, Bellagio, Italy, June 2005.

2   These estimates are obtained by comparing household expenditures with official poverty lines. Because of a change in survey design, poverty estimates of 1999 are not strictly comparable to earlier poverty estimates. Various researchers have produced 'adjusted' estimates – the number reported in the text is on the higher side of these estimates (Kijima and Lanjouw, 2005).

3   In the 1980s, associated with the diffusion of high yielding seeds, crop output grew at more than 3 per cent per annum compared to about 2.3 per cent earlier. In the period since, growth rates have slumped back to 2.2 per cent per annum.

4   See Herring (2005) for an analysis of how opposition to GM crops has been constructed and how it has played out in politics.

5   The involvement of the Ministry of Health has been marginal. This could

change with the debate about labelling norms and laws for GM foods.

6   The experience of JK Seeds and Nath Seeds – domestic private seed companies that are developing *Bt* cotton hybrids with non-Monsanto genes – will be instructive in this regard.

7   A packet consists of 450 grams of seed.

8   Although India now allows patents to biotechnology innovations, it may well keep in public domain the key elements of genomics and the basic biotech tools. It is not clear, therefore, that patents will hamper public sector research.

References

Bambawale, O. M., Singh, A., Sharma, O. P., Bhosle, B. B., Lavekar, R. C., Dhandapani, A., Kanwar, V., Tanwar, R. K., Rathod, K. S., Patange, N. R. and Pawar, V. M. (2004), 'Performance of *Bt* cotton (MECH-162) under Integrated Pest Management in farmers' participatory field trial in Nanded district, Central India', *Current Science*, vol 86 (12), pp1628–1633

Bennett, R. M., Ismael, Y., Kambhampati, U. and Morse, S. (2004) 'Economic impact of genetically modified cotton in India', *AgBioForum*, vol 7 (3), pp96–100, www.agbioforum.org

Buttel, F., Merrill, J., Chen, L., Goldberger, J. and Hurley, T. (2005) 'Bt corn farmer compliance with insect resistance management requirements', Staff Paper P05-06, Department of Applied Economics, University of Minnesota.

Datt, G. and Ravallion, M. (1998) 'Farm productivity and rural poverty in India', *Journal of Development Studies*, vol 34, pp62–85

Eswaran, M. and Kotwal, A. (1993) 'A theory of real wage growth in LDCs', *Journal of Development Economics*, vol 42, pp243–270

Eswaran, M., Kotwal, A., Ramaswami, B. and Wadhwa, W. (2006) 'How poverty declines: Lessons from India', Indian Statistical Institute, Delhi, mimeo

GoI (Government of India) (2004) *Report of the Agricultural Biotechnology Task Force*, Ministry of Agriculture, New Delhi

GoI (2005) *National Biotechnology Development Strategy*, draft, Department of Biotechnology, Ministry of Science and Technology, New Delhi

Grover, A. and Pental. D. (2003) 'Breeding objectives and requirements for producing transgenics for major field crops of India', *Current Science*, vol 84 (3), pp310–320

Grover, A., Aggarwal, P. K., Kapoor, A., Katiyar-Agarwal, S., Agarwal, M. and Chandramouli, A. (2003) 'Addressing abiotic stresses in agriculture through transgenic technology', *Current Science*, vol 84 (3), pp355–367

Herring, R. J. (2005) 'Miracle seeds, suicide seeds and the poor: NGOs, GMOs and farmers', in Ray, R. and Katzenstein, M. (eds) *Social Movements in India: Poverty, Power, and Politics*, Rowman and Littlefield, Lanham, Md, Oxford University Press, New Delhi

Jain, S. (2001) 'Bt in *Bt* cotton means Blocking the seed, Trashing the fact', *Indian Express*, 27 October 2001

Jain, S. (2002) 'The IT revolution missed Punjab, we won't let *Bt* pass us by', *Indian Express*, 9 March 2002

Kijima, Y. and Lanjouw, P. (2005) 'Economic diversification and poverty in rural India', *Indian Journal of Labour Economics*, vol 48 (2), pp349–374

Kremer, M. and Zwane, A. P. (2005) 'Encouraging private sector research for tropical agriculture', *World Development*, 33 (1), pp87–105

Lalitha, N., Ramaswami, B. and Pray, C. E. (2006) 'The limits of intellectual property rights: Biosafety regulation and illegal seeds in India', mimeo, Planning Unit, Indian Statistical Institute, New Delhi

Lipton, M. and Longhurst, R. (1989) *New Seeds and Poor People*, Heritage Publishers, New Delhi

Murugkar, M., Ramaswami, B. and Shelar, M. (2006) 'Liberalization, biotechnology and the private seed sector: The case of India's Cotton Seed Market', discussion paper no 06-05, Planning Unit, Indian Statistical Institute, New Delhi, available at www.isid.ac.in/~planning/workingpapers/dp06-05.pdf

Naik, G., Qaim, M., Subramanian, A. and Zilberman, D. (2005) 'Bt cotton controversy: Some paradoxes explained', *Economic and Political Weekly*, XL, 15, pp1514–1517

Nuffield Council on Bioethics (2004) *The Use of Genetically Modified Crops in Developing Countries*, London, www.nuffieldbioethics.com

Pental, D. (2005) 'Transgenic crops for Indian agriculture: An assessment of their relevance and effective use', in Chand, R. (ed) *India's Agricultural Challenges*, Centad, New Delhi, available at www.centad.org/publications_1.asp

Pray, C., Bengali, P. and Ramaswami, B. (2005) 'The cost of bio-safety regulation: The Indian experience', *Quarterly Journal of Agriculture*, vol 44 (3), pp267–289

Qaim, M. (2003) 'Bt cotton in India: Field trial results and economic projection', *World Development*, vol 31 (12), pp2115–2126

Rai, M. (2006) 'Harnessing genic power to enhance agricultural productivity, profitability and resource use efficiency', Twelfth Dr B. P. Pal Memorial Lecture, New Delhi

Ravallion, M. and Datt, G. (1996) 'How important to India's poor is the sectoral composition of economic growth', *World Bank Economic Review*, 10 (1), pp1–25

Sahai, S. and Rehman, S. (2004) '*Bt* cotton 2003–2004, fields swamped with illegal variants', *Economic and Political Weekly*, vol 39 (24), pp2673–2674

Sharma, M., Charak, K. S. and Ramanaiah, T. V. (2003) 'Agricultural biotechnology research in India: Status and policies', *Current Science*, vol 84 (3), pp297–302

Srinivasan, C. S. (2004) 'Plant variety protection in developing countries: A view from the private seed industry in India', *Journal of New Seeds*, vol 6 (1), pp67–89

10

South Africa: Revealing the Potential and Obstacles, the Private Sector Model and Reaching the Traditional Sector

Marnus Gouse

Introduction

South Africa (SA) has a proud and solid biotechnology R&D history. Some of this R&D has lead to the establishment of one of the largest beer breweries in the world, production of world class wines, creation of new animal breeds and plant varieties that are used all over the world, and competitive domestic dairy and yeast industries (DACST, 1996). However, in relation to developed and some developing countries, the country has failed to benefit from the emergence of 'modern' biotechnology.

Over the last 25 years SA has been developing genetic engineering techniques and capacity and over 600 research projects are now underway. But only a small number of marketable products and techniques have been developed. Reasons – and possible solutions – for this are discussed in detail below. But first, the history of GM crops in SA and the structure of the SA agricultural sector are summarized in the first two sections. Farm level income and yield effects of GM crop adoption are then discussed in the third section. The last three sections identify some institutional challenges for commercial dissemination, shed light on the development and challenges of the biosafety regulatory system, and set forth some policy challenges.

History and background

History of agricultural biotechnology and genetically modified crops in South Africa

An application from the US seed company, Delta and Pine Land in 1989 to perform field trials of genetically modified cotton kick-started the South African biosafety process and initiated the first trials with GM crops on the African continent. D&PL used South Africa as an over-wintering haven for field trials.

In 1997, SA became the first country in Africa to commercially produce GM crops. To date the commercial release of insect-resistant *Bt* cotton and maize, as well as herbicide-tolerant RR soybeans, cotton and maize have been

Table 10.1 *Percentage and estimated areas planted to GM crops in South Africa*

Crop	1999/2000	2000/2001	2001/2002	2002/2003	2003/2004
% *Bt* cotton	50%	40%	70%	70%	81%
Bt cotton area (ha)	13,200	12,000	25,000	18000	30,000
% RR cotton	0	0	<10%	12%	7%
RR Cotton area (ha)	0	0	1500	3500	2500
% *Bt* yellow maize	3%	5%	14%	20%	27%
Bt yellow maize area (ha)	50,000	75,000	160,000	197,000	250,000
% *Bt* White Maize	0	0	0.4%	2.8%	8%
Bt white maize area (ha)	0	0	6000	55,000	175,000
% RR soybeans	0	0	5%	10.9%	35%
RR soybeans area (ha)	0	0	6000	11,000	47,000

Sources: Percentages from Cotton SA and South African Seed Organization; area from author's own estimations

approved. Cotton with the 'stacked gene' (*Bt* and **RR**) was approved in 2005 after a review period of close to three years. Farmers started adopting insect resistant *Bt* cotton varieties in the 1997/1998 season and insect resistant *Bt* yellow maize in the 1998/1999 season. Herbicide tolerant cotton was made available for commercial production in the 2001/2002 season and a limited quantity of herbicide tolerant soybean seed was also released. *Bt* white maize was introduced in the 2001/2002 season and 2002/2003 saw the first season of large-scale *Bt* white maize production. A limited quantity of herbicide tolerant maize seed was commercially released for the 2003/2004 season. Table 10.1 summarizes the areas planted to GM crops in South Africa for the most recent seasons.

The supplier of these varieties is D&PL, which has the sole use of Monsanto's *Bt* cotton gene in its cotton seed in the country. In 1997/1998 D&PL introduced two *Bt* varieties that were not initially adopted with great enthusiasm. In 1998/1999 D&PL's cotton seed market share was estimated at around 10 per cent, and increased to close to 20 per cent in 1999/2000. When NuOpal (Opal with *Bt*) was introduced for the 2000/2001 season, that market share soared to over 80 per cent and with the release of herbicide tolerant cotton in 2001/2002, to over 95 per cent (Gouse et al, 2005c).

As can be seen in Table 10.1, the initial spread of *Bt* yellow maize was quite slow, reaching only 3 per cent of the total maize area after two years. Farmers and seed companies suggested three reasons for this. First, the *Bt* hybrids initially on the market were well adapted to neither local consumer markets nor local agricultural production conditions. Second, in the initial adoption seasons many farmers did not see a large productivity advantage because stalk borer pressure was relatively low. In addition, farmers felt that the increased yield from *Bt* maize did not justify the additional technology fee, especially when compared to yields from newer, better adapted conventional hybrids. Thus, at first *Bt* was most likely to be adopted mainly in those areas where stem borers were a difficult problem. Third, farmers worried that they might not be able to sell their crops because of consumer concerns about GM food.

By 2000/2001, seed companies in SA had been able to cross the *Bt* gene into more appropriate local yellow and white maize hybrids. In 2001/2002, there was a significant stalk borer attack; farmers suffered damage and subsequent yield losses on their conventional maize and *Bt* maize rendered significant yield benefits. This changed farmer perceptions. The final adoption constraint – consumer acceptance – has thus far not become a significant reality. In the majority of cases farmers have had no problem in selling their GM maize and a price premium on non-GM maize has been more the exception than the rule. Non-GM maize has been exported to niche markets, mainly in Asia, but the extra profits were essentially captured by the commodity trading companies and not the producers.

Agricultural sector: Dualistic structure and liberalization reforms

Agriculture plays an essential role in the South African economy by contributing to food security, earning foreign currency and providing a vital source of employment. Even though agricultural production's contribution to GDP has been declining over the last decades, it still accounts for about 4.5 per cent of GDP. Investment by the previous 'apartheid' government has translated into the geographically inequitable establishment of key support services such as roads, railways, communication links, agricultural training centres, dams and irrigation facilities, input distribution outlets, and research, extension and financial services. Between 1985 and 1993, budgetary allocations to commercial agriculture averaged 67 per cent of the total agricultural budget, compared to 33 per cent for all the former homelands combined (DBSA, 1993). About 46,000 large-scale, mainly white commercial farmers occupy 87 per cent of the total agricultural land, produce the bulk of agricultural production and guarantee national food security. Commercial agriculture contributes 8.4 per cent of SA's total export earnings (Kirsten, 2006). The commercial farmers employ approximately 1 million workers (equivalent to 10 per cent of formal sector employment) and provide housing, schooling and livelihoods for an estimated 6 million family members of these workers. On the other side of the economic divide an estimated 240,000 small-scale farmers support over 1 million family members and provide occasional employment for about 500,000 people. It is estimated that an additional 2.5 million farmers in South Africa produce mainly white maize and vegetables on a subsistence level. Thus, it is important to take both small- and large-scale farmers into consideration when new policies are formulated and agricultural technologies introduced.

For 60 years agriculture was highly protected. The early 1980s saw a decline in the use of price controls and a shift to market-based pricing systems. General Agreement on Tariffs and Trade (GATT) negotiations enhanced pressure to abolish quantitative import controls and tariffs on agricultural commodities. Between 1991 and 1996, ten agricultural marketing boards were abolished and the Marketing of Agricultural Products Act of 1996 set out to prevent, rather than promote, government interventions with the main objectives of providing equitable market access and promoting efficiency in marketing of agricultural products (Meyer, 2002). Farmers suddenly had to compete in an open economy with world prices (deflated due to subsidies of developed countries) influencing domestic crop prices.

The vast majority of crops are dependent upon rainfall, which is generally low and highly variable from year to year and through the production season. Production risk, combined with market liberalization and other factors like a volatile exchange rate, new labour standards, minimum wages, land reform (restitution and redistribution), a new Water Act controlling water access/usage and Agricultural Black Economic Empowerment (AgBEE) have put South African farmers under immense pressure to survive. As they search for ways to decrease production risk and increase efficiency and profitability, the impressive adoption rates of GM crops (see Table 10.1) are not surprising.

Farm level yield and income impacts for small-scale farmers in South Africa

Although commercial farms account for most of the area under GM cotton and maize production, it is the experience of the small-scale and subsistence farmers in South Africa that has received more media and research coverage. The performance of *Bt* cotton on the Makhatini Flats has been hailed by industry and pro-biotech bodies as the first real example of how GM crops can assist resource poor farmers and better the life of rural households in developing countries. In 2001, SA became the first country in the world where a GM staple crop had been produced by subsistence farmers.

Small-scale farmers and *Bt* cotton: Makhatini Flats experience

Studies on the impact of *Bt* cotton adoption among small-scale farmers in SA focused on the Makhatini Flats in northern KwaZulu Natal (KZN), the larger of only two areas in SA where smallholders have fairly continuously produced cotton over the last two decades. When Monsanto received permission to sell *Bt* cotton seed for the 1997/1998 cotton production season, efforts were made to plant demonstration plots on the Makhatini Flats in order to show that the technology is scale neutral. According to Monsanto, the results were impressive, and the following year 75 small-scale farmers planted *Bt* cotton on approximately 200 hectares. In 1999/2000 the number of *Bt* adopters rose to 411 farmers on about 700 hectares, and in 2000/2001 to 1184 adopters on about 1900 hectares (Bennett, 2002). This rapid adoption rate can be attributed partly to the success of the farmers who first adopted the technology but also to the input and extension supplier and cotton buyer on the Flats, Vunisa (meaning 'to harvest' in isiZulu), which recognized the fact that *Bt* outperformed conventional varieties and thus recommended the new seed to farmers in order to decrease its credit risk. Up until the 2004/2005 production season, more than 90 per cent of the cotton on the Flats was GM. But in the following season there was a dramatic decline in the total cotton area. The reasons for this were lack of production credit and the low seed cotton price, as explained below.

A number of studies showed that farmers on the Flats benefited from *Bt* cotton adoption, mainly through increased yields and fewer pesticide applications. The ginning company (Vunisa), in partnership with the Land Bank of South Africa, was supplying production credit and in 1998/1999 there was a loan recovery rate of close to 90 per cent. Then in 2001/2002 a new ginning company erected a gin on the Flats. A substantial number of farmers who had taken production credit from Vunisa avoided repaying their loans by selling their cotton to the new gin. Having suffered substantial losses, Vunisa and the Land Bank no longer offered loans in 2002/2003 and in the 2004/2005 production season there was still no credit available to small-scale cotton farmers on the Makhatini Flats. With the situation exacerbated by a low seed cotton price, the area planted to cotton on the Flats decreased drastically. Anti-GM campaigners have pointed to this fall in production area and drop in number of farmers as proof that GM cotton has failed and even plunged Makhatini farmers into a debt crisis.

The truth, however, is that the majority of farmers on the Flats cannot finance their own cotton production inputs. A study by the French research institute, Agricultural Research Centre of International Development (CIRAD) (Hofs et al, 2005), showed that the farmers on the Flats who were able to produce during the seasons when credit was not available were mainly the elderly who could finance inputs with pension money. The vast majority of these self-financing farmers still bought GM cotton. It is also true, however, that due to credit (i.e. grants) supplied in the past by the SA government through development programmes in the former homeland areas, and inability of financial corporations to enforce repayment of loans, adverse selection by farmers on the Flats is not uncommon. Borrowing money from one gin and supplying the cotton to another gin is hardly a new or rare occurrence in Africa, where small-scale landownership is the exception rather that the rule and the only asset that can be used as collateral for a loan is the potential harvest. Thus, in a periodically drought-stricken country like South Africa, it does happen that farmers are not able to repay loans even if they deliver all of their cotton. This might tempt a farmer to deliver his cotton under a different name, or to break a contract and deliver his harvest to a different buyer. It is also possible that even though a farmer's gross margin may have increased due to *Bt* cotton adoption, he/she might still not see a profit at the end of the season. The Makhatini Flats cotton story emphasizes that without good governance and institutional structures, the potential gains of GM crops will not be realized (Gouse et al, 2005a).

Studies show that small-scale farmers enjoyed larger benefits than large-scale commercial farmers. Gouse et al (2003) found an 18.5 per cent yield increase for large-scale irrigation farmers for the 2000/2001 season, which compares well with a 16.8 per cent increase measured on field trials at the Clark Cotton experimental farm in Mpumalanga. Large-scale dryland farmers enjoyed a 14 per cent yield increase while small-scale dryland farmers enjoyed an increase of between 16 per cent and above 40 per cent in 1998/1999 and 1999/2000 (Gouse et al, 2005a). These trends are consistent with findings elsewhere, such as in Argentina (Qaim et al, 2003), where large-scale commercial farmers were reported to enjoy 19 per cent yield increases and small-scale farmers had 41 per cent yield increases. Like Argentina studies, the South African researchers attribute the difference between the *Bt* yield advantages of small- and large-scale farmers to the financial and human capital constraints that cause smallholders to invest in chemical pest control. Shankar and Thirtle (2005) showed that the average insecticide application level of small-scale farmers on the Flats is lower than 50 per cent of the optimal level. Many small-scale farmers indicated that they were not even able to apply pesticides on their whole field due to a lack of time, knapsack sprayers, labour and the cost of pesticide. With a low education level causing problems with the mixing of pesticides and the calibration of knapsack spraying nozzles, the efficacy and efficiency of insecticide applications is questionable for a large number of small-scale farmers.

Small-scale farmers also enjoyed larger financial benefits because they paid a substantially lower technology fee; R230 (US$29) for a 25-kilogram bag of seed compared with R600 (US$75). This lower fee can be explained by a combination of possible factors including willingness to pay and an effort towards poverty alleviation by the multinational technology innovator (something not unheard of in Africa). But more likely it resulted from an endeavour to establish a market for GM cotton among small-scale producers as the smallholder farming conditions in South Africa are more applicable to the rest of Africa than those of South Africa's large-scale farmers. Another possible reason is that selling cotton seed through Vunisa made it possible to supply *Bt* seed to small-scale cotton farmers at lower prices than to large-scale farmers. The Vunisa depot on the Flats served only small-scale farmers and there were no large-scale farmers active in the area.

Gouse et al (2005c) showed that despite facing a monopolist seed supplier and a monopolist gene supplier in 1999/2000 and 2000/2001, South African farmers captured the lion's share of the additional welfare created by the introduction of *Bt* technology in the South African cotton sector (see Table 10.2). Even though D&PL's share of the additional profit seems small, through their agreement with Monsanto they were able to secure most of the South African cotton seed market. The technology supplier captures its share through the technology fee and farmers benefit through increased yields and savings on insecticides. Other benefits indicated by farmers included peace of mind, managerial freedom and savings on spraying, water, labour and machinery, but these were not quantified or included in the calculation. Based on the 1999/2000 technology fees, calculations show that if a small-scale farmer had to pay the

Table 10.2 *Distribution of additional benefit according to farmer type*

	Small-scale Dryland Farmers	Large-scale Dryland Farmers	Large-scale Irrigation Farmers
Seed company: D&PL	R32,500 3%	R54,600 2%	R74,200 1%
Technology supplier: Monsanto	R299,400 28%	R1,309,800 52%	R1,779,700 20%
Farmer	R1,038,600 69%	R1,323,500 45%	R5,988,100 79%
Primary consumer: Ginning company	0%	0%	0%
Insecticide companies	R90,600	R777,700	R1,086,400

Note: US$1=R8.

same technology fee as a large-scale farmer, the small-scale farmer's share of the benefit would decrease to approximately 46 per cent (from 69 per cent), while 52 per cent (increasing from 28 per cent) would accrue to Monsanto. Currently all farmers are paying the same technology fee based on whether cotton is produced on dryland areas or with irrigation.

Subsistence farmers: *Bt* maize experience

In 2001/2002, South Africa became the first country in the world to permit the commercial production of a GM subsistence crop, i.e. insect-resistant white maize. In South Africa and other Southern African countries, the losses sustained in maize crops due to damage caused by the African maize stem borer (*Busseola fusca*) are estimated to be between 5 per cent and 75 per cent and it is generally accepted that African maize borer annually reduces the South African maize crop by an average of 10 per cent (Annecke and Moran, 1982). Both *Busseola fusca* and *chilo partellus* can be controlled to a satisfactory level with the use of the *Bt* gene currently used in South African *Bt* varieties, *Cry1Ac*.

Gouse et al (2005b) found that despite paying more for seeds, commercial farmers who adopted *Bt* maize enjoyed increased income from *Bt* maize compared to conventional maize through savings on pesticides and increased yield due to better pest control. A study by the University of Pretoria found statistically significant yield increases of around 11 per cent with *Bt* yellow maize and farmers in the Douglas irrigation scheme area (Northern Cape) report yield increases of approximately 2 tonnes (about 12 per cent) with *Bt* maize when stalk borer pressure is high.[1]

In 2001/2002, Monsanto introduced *Bt* white maize to small-scale farmers through workshops in nine areas in four provinces in South Africa. Farmers who wished to try out the new seeds received two small bags of white maize seed. One of the bags contained 250 grams of Yieldgard or insect-resistant *Bt* maize seed, while the other bag contained 250 grams of the isoline conventional variety.

Gouse and Kirsten (2004), studied the first three seasons of this experience. The first season rendered some interesting results, with 175 small-scale farmers reporting yield increases averaging 32 per cent. The 2002/2003 season saw an impressive demand for *Bt* seed from various sites (the Transkei area, for instance, ordered 4.5 tonnes of *Bt* seed), but in only two sites in KZN were a significant number of subsistence farmers able to purchase *Bt* white maize seed.

For a large number of small-scale/subsistence farmers in different areas in South Africa, sourcing specific seed varieties is much more difficult than for large-scale farmers. Large-scale maize farmers order their maize seed months in advance, in most cases directly from the seed companies. Seed can also be bought from large, local agricultural input suppliers (former farmer cooperatives). Small-scale farmers and subsistence farmers in South Africa who produce maize on a larger scale than just a vegetable garden are mainly situated in former homelands, such as the formerly independent Transkei and Ciskei areas in the Eastern Cape and designated tribal areas like Zululand in Northern KZN. There are few maize producing large-scale farmers active in these areas and economies of scale (relatively small quantities, packaged in small units and transported over long distances) cause problems for input suppliers. In some areas farmers, supported by provincial department of agriculture offices, have formed small cooperatives to buy inputs in bulk and to negotiate better prices, but in many of these rural areas small-scale farmers are still dependent on the local general store to order some type of maize seed in time for planting.

Despite a lower than normal rainfall and stalk borer pressure in 2002/2003, small-scale farmers in KZN enjoyed a statistically significant (95 per cent level) yield increase of 16 per cent due to better stalk borer control with *Bt* maize. These farmers were better off than farmers who planted conventional hybrids, despite the additional technology fee. In 2003/2004 no significant difference between the yields of *Bt* and conventional maize seed could be found due to drought, a very low stalk borer infestation level and damage to maize ears caused by late rain, which complicated measuring yield comparison. (As with large-scale farmers, the benefits of *Bt* maize for small-scale farmers depend on the presence of stalk borers in a particular season.) In the third season, which was also the fourth consecutive dryer than usual season in the research area, the stalk borer infestation level was very low and farmers who planted *Bt* maize enjoyed yields similar to those of farmers who planted conventional hybrids. Thus, in all likelihood, the former were marginally worse off due to the additional technology fee. It seems that small-scale farmers have realized this to a certain extent, not only for *Bt* maize seed but also for expensive conventional hybrids, because in the 2003/2004 and 2004/2005 seasons a large number of farmers in KZN bought seeds for *Bt* and conventional hybrids but decided not to plant them after a very dry pre-season. In order to minimize financial risk they planted traditional seed or less expensive open-pollinated varieties without fertilizer, planning to plant the *Bt* and conventional hybrids with fertilizer the next season, which they hoped would be wetter and more maize production friendly.

R&D in the public and private sectors

Institutional challenges for R&D

A 2003 National Biotechnology Survey (DST, 2004) found 622 research groups engaged in 911 research projects relevant to biotechnology, spread over human and animal health, plants, foods and beverages, and industrial, environmental and other fields. The human health sector has the most projects with the plant sector in second place. The majority of the research groups engaged ten or fewer researchers.

The survey also identified 106 companies active in modern biotechnology, of which 47 use biotechnology as a main business focus. The firms are small: fewer than 20 per cent have an annual revenue above R10 million (US$1.5 million). It is estimated that only 10% of them are conducting innovative cutting edge research and development, with the majority involved in new applications of low-tech modern biotechnology (DST, 2004).

The private sector performs just over half of national R&D, the tertiary education sector a quarter, and the government about 20 per cent. R&D expenditures by the private and public sectors are approximately equal (Wolson, 2005a). Financing is a major constraint in all sectors. In the private sector, a relatively risk-averse venture capital sector has until recently displayed little interest in investing in biotechnology and the general environment for commercialization of biotechnology has not been conducive to the establishment of private biotechnology firms. In the public sector, despite its recognition of the contribution and importance of agriculture, in the latter part of the 1990s, the new government decreased expenditure on R&D, focusing on new national and political priorities and endeavouring to rectify the inequalities of the past. Gross domestic expenditure on R&D is estimated at about US$1.5 billion (adjusted for PPP) and amounts to 0.76 per cent of GDP. The government plans to increase public spending on R&D to 1 per cent of GDP over the next couple of years but it will still be much lower than the OECD average of 2.5 per cent (Wolson, 2005a).

R&D capacity of the National Department of Agriculture is very limited and focuses mainly on adaptive agricultural research. The Agricultural Research Council (ARC) is the largest agricultural research entity in South Africa, accounting for nearly 60 per cent of the country's agricultural R&D expenditure and researchers in 2000 (Liebenberg et al, 2004).

Lack of expertise and skilled personnel is the major constraint to the development of biotechnology (DACST, 2001). Wolson (2005a) also suggests that institutional arrangements have not promoted sufficient linkages between researchers in different disciplines and/or organizations, or among researchers, industry and government. There are also too few researchers in the labour force due to limited employment opportunities for graduates in the local biotechnology industry and much better opportunities overseas (in SA an average post-doctoral income is estimated to be only 40 per cent that earned by a post-doctoral fellow abroad (DACST, 2001)). The 'brain drain' has consequently left limited numbers of skilled people for entrepreneurship in biotechnology.

Despite these systemic constraints, a number of South African public institutions have been able to achieve a number of biotechnology applications in recent years. For example:

- Three ARC institutes are developing GM applications in potatoes, tomatoes, maize, lupins, soybeans, tobacco, melons, pears, apples, apricots, plums, strawberries and groundnuts. Field trials have been conducted using the ARC's own research materials, as well as under contract with private companies, testing maize, cotton, tomatoes, potatoes and strawberries. Infruitec developed a herbicide resistant strawberry. Though results were extremely good, the project was shelved because the private chemical company involved found it too expensive to license the use of the herbicide on strawberries in South Africa.

- The Forestry and Agricultural Biotechnology Institute at the University of Pretoria has been pursuing research on cereal genomics, fungal genetics, molecular plant physiology, citrus, banana and mango, molecular plant-pathogen interactions and other forest molecular genetics. Its flagship Tree Protection Cooperative Programme, focusing on tree disease problems, is a cooperative research venture among the major players in the forestry industry.

- The Department of Molecular and Cell Biology at the University of Cape Town is developing maize resistant to the African maize streak virus and maize tolerant to drought and other abiotic stresses. It is also investigating use of tobacco plants to produce vaccines for human use.

- The South African Sugar Experimental Station has developed herbicide tolerant sugarcane, now in the field trial stage.

- The Council for Scientific and Industrial Research (CSIR) has genetically engineered maize with a gene isolated from beans to develop resistance to the fungal pathogen, *Stenocarpella maydis*. This project is at the field trial stage. Also, in collaboration with overseas partners, the CSIR has introduced four antifungal genes into maize to confer resistance to *Fusarium moniliforme*. Pearl millet has been engineered to render it resistant to *Sclerospora graminicola*, which causes downy mildew (Thomson, 2004).

However, most of these results have remained at the laboratory level. While a few have gone to field trial, none has gone on to development of a commercial variety. Most of the research was done as proof of concept and without substantial funding to support the comprehensive and long-term process of developing a real marketable product; these applications stayed on the shelf.

International collaboration and public–private partnerships are an important part of some research programmes. Using the *(Bt)-cry1la1* gene obtained from Syngenta, the ARC's Roodeplaat Vegetable and Ornamental Plant Institute is developing an insect-resistant potato in collaboration with Michigan State University and USAID. Supported by the Grand Challenges in Global Health initiative, primarily funded by the Bill & Melinda Gates Foundation, the CSIR is involved with the Africa Biofortified Sorghum project, which aims to develop more nutritious, easily digestible sorghum varieties containing greater levels of vitamins and protein. Some of the other eight organizations involved in this project include the ARC Grain Crops Institute, the University of Pretoria, the University of Missouri-Columbia in the US, Africa Harvest and DuPont, through its Pioneer Hi-Bred International seed company.

The main multinational agricultural biotechnology and seed companies in South Africa (Monsanto, Syngenta and Pioneer Hi-Bred) have not set out to do groundbreaking biotech research in the country. Rather, they have introduced their already developed technologies or acquired events into locally developed germplasm and are focusing their resources on developing improved varieties and hybrids, making use of marker-assisted breeding. Monsanto not only sells its *Bt* gene in its own hybrids but also licenses this gene to other maize seed companies in South Africa, including a local seed company, Pannar, for use in its hybrids. In 2003, the biosafety committee of SA approved Syngenta's *Bt* maize, breaking Monsanto's monopoly on *Bt* genes in SA.

Pannar has recently invested in a new biotechnology centre at its head office in Greytown, KZN. Activities will focus on monitoring genetic quality through the plant breeding and seed production processes and ensuring quicker and more targeted plant breeding through the use of gene technologies such as marker-assisted breeding. The centre will not only support Pannar Seed's plant breeding programmes in SA but will also work closely with Pannar's research operations in the US, Argentina and Europe and will be used extensively to enhance the vegetable breeding capabilities of Pannar's sister company, Stark Ayres.

Policy initiatives to encourage biotechnology R&D

In The White Paper on Science and Technology of 1996 the South African government indicated that it considers science and technology central to creating wealth and improving the quality of life in contemporary society. It recognized its responsibility for creating an enabling environment for innovation, specifically as a means of achieving the national imperatives of reducing the impact of HIV/AIDS, job creation, rural development, urban renewal, crime prevention, human resource development and regional integration. It is believed that biotechnology can play a major role in addressing these objectives (DACST, 1996).

Realizing that South Africa has failed to optimally extract value from the recent advances in biotechnology the government released a National Biotechnology Strategy for South Africa in 2001. The document (Government of South Africa, 2001) again identifies biotechnology as a key technology platform from which to address the above-mentioned national imperatives and announces a series of policy initiatives to stimulate the development of an active and productive South African biotechnology industry. The government committed R450 million (approximately US$75 million) from 2004 to 2007 to fund these initiatives. To increase specific focus, the Department of Arts, Culture, Science and Technology was divided into the Department of Science and Technology and the Department of Arts and Culture in August 2002.

Several initiatives have since emerged. First was the Biotechnology Regional Innovation Centres (BRICs) that facilitate sharing of capital, equipment and specialized expertise (Wolson, 2005b). Three BRICs have been set up, focusing on diverse aspects and areas of applications from human health to industry to crops and livestock. Another initiative was creation of the National

Innovation Centre for Plant Biotechnology (PlantBio), focusing on all aspects of plant biotechnology.

In early 2003, the South African Agency of Science and Technology Advancement launched the programme Public Understanding of Biotechnology (PUB) to promote broad public awareness and a clear understanding of the potential of biotechnology (in food and agriculture) and to stimulate dialogue and debate on its current and potential future applications, including genetic modification. The target audience includes all facets of society with emphasis on consumers, educators and learners. In a recently released study commissioned under the programme, it was found that South Africa's knowledge and understanding of biotechnology is limited. In reply to the question 'What do you think when you hear the word biotechnology?', 82 per cent of 7000 respondents indicated that they did not know what they thought (HSRC, 2005).

Another initiative, the National Bioinformatics Network (NBN), was created to develop capacity in bioinformatics in South Africa, especially among previously disadvantaged groups. Its vision is to bring South Africa into the mainstream of this field within ten years, making it a leader among developing countries.

In 2003, a biotechnology incubation institute, eGoli BIO Life Sciences Incubator, was launched to serve as a development conduit for the commercialization of life sciences research, products, services and technology platforms by supplying business infrastructure, strategic guidance, financial and legal advice, and by creating a learning/sharing environment in which information, experience and ideas are freely exchanged. A number of new companies have made use of this initiative, which currently has four tenant firms.

Biosafety regulation

In 1978, the South African Committee for Genetic Experimentation was formed to be responsible for recovering all aspects of recombinant DNA, provide guidelines, and approve and classify research centres and projects and advise the National Department of Agriculture. This system was succeeded by the provision of the 1997 Genetically Modified Organisms Act that established three new structures to regulate all aspects of GM crops: the Executive Council (the national, independent decision-making body that includes representatives of Departments of Agriculture, Environment, Tourism, Health, Trade and Industry, Labour, and Science and Technology); the Scientific Advisory Committee; and the Registrar and Inspectorate in the Ministry of Agriculture that monitors local implementation.

GM crop regulations stipulate that the process of approval or rejection of an application should not take longer than 90 days for a decision on field trials and 180 days for a decision on general release applications (Thomson, 2002). In the period 2000 to 2004, more than 900 permits for glasshouse and field trials, contained use, commodity clearance, imports and exports were reviewed and granted. The applications for commercial release of insect resistant cotton and maize and herbicide tolerant cotton, maize and soybeans were reviewed and approved in what has been regarded internationally as a scientifically responsible yet efficient manner.

Intellectual property rights

SA has well developed legislation on intellectual property rights (IPR), patent rights and plant breeders' rights and a respected legal and judiciary system to enforce compliance. The country has ratified a number of the relevant international property rights treaties and conventions including TRIPS, Convention on International Trade in Endangered Species of Wild Fauna and Flora (CITES), Commission on Genetic Resources for Food and Agriculture (CGRFA) and the Cartagena Protocol on Biosafety. Several other acts have been passed to facilitate ongoing availability of quality plant genetic resources so as to support production efficiency in agriculture, an orderly industry and an environment conducive to R&D (Van Der Walt and Koster, 2005).

Under the Patent Act, the patenting of plants and animals or natural resources is excluded. GM varieties are usually protected under a patent on the genetic construct and claims associated with the modification, a trademark on the name and often a plant breeder's right for the specific variety. Sale of GM seed is often accompanied by a contract stipulating that farmers shall not retain seed for distribution or sales purposes.

The government is currently drafting legislation for a revised IPR framework. It is expected that, among other things, the review will examine the impact that the current IPR framework has on development goals and provide more incentives to research. It has been suggested that the new legislation might roughly follow the US Bayh-Dole Act, in which research institutions are granted ownership of inventions developed with government funds, but it would also have to take into account the relative youth and small size of the biotechnology sector.[2]

Technology suppliers have had no problem with GM maize and cotton farmers contravening the terms and conditions regarding the replanting of harvested seeds. Maize farmers in South Africa who buy and plant hybrid seed are used to buying seed each season in their quest to obtain the best quality and highest yield-potential seed possible. Seed cotton is delivered into a closed cycle where the seed is delinted by a cotton gin and then delivered to the seed supplier or into the animal feed supply chain. Farmers are allowed to replant their harvested seeds on their own farms but for them to try to delint and clean cotton seed with a specialized acid treatment in order to make it clean enough for planting with a mechanical planter is hardly worth the effort and also runs the risk of damaging seeds. So South African cotton farmers also have to buy seed every season.

Governing soybean replanting is more difficult. Farmers are allowed to replant their own harvested soybean seed but, as with cotton, not allowed to sell or distribute it to neighbours since the seed variety is protected by plant breeders' rights and the technology trait by property rights. It is estimated that between 60 per cent and 80 per cent of the soybeans in SA are produced from replanted seeds.[3] Since the release of RR soybeans in SA, farmers do not pay a technology fee when purchasing seed of GM varieties, but only if their harvest tests positive when it is delivered to a silo. In 2005, a farmer delivering RR soybeans was charged R65 (approximately US$10) per tonne. This system is not ideal, as not all farmers deliver their harvest to an official silo or buyer, but it is probably less arduous than policing every soybean farmer at planting time. South Africa's climate is not well suited to soybean production and as a net importer, it produces only about 120,000 hectares of the crop annually, depending on the season and the price of maize.

Marketing of seeds and products

The seed market

South Africa has a well-developed seed sector with an annual turnover close to R1000 million (US$160 million), of which maize is a major crop. The South African variety list of October 2004 had more than 400 registered white and yellow maize hybrids available for commercial production. These hybrids are sold by 18 companies but four companies dominate the market. The South African company Pannar owns 44 per cent of the hybrids on the variety list, Monsanto 23 per cent, Pioneer Hi-bred 18 per cent and Syngenta 16 per cent. Monsanto, through its acquisition in 1999/2000 of two South African seed companies (Carnia and Sensako) has been able to capture a major share of the maize seed market and also owns 48 per cent of the wheat hybrids on the variety list. All these companies' germplasm R&D has mainly focused on developing high yielding varieties for commercial farmers using plenty of fertilizer. Only the ARC and Pannar have purposefully invested in development and marketing of maize varieties more conducive to subsistence farming. Syngenta (with its partnership with Seedco), which recently relocated to Zambia from Zimbabwe, might also be heading in this direction.

The truth is that there have been few constraints to the commercial dissemination of GM crops in South Africa. With plant breeders' rights, intellectual property rights, a well-developed seed sector and infrastructure and the biosafety regulatory system in place, the introduction and commercialization of GM crops in South Africa has been smooth, commercially safe and straightforward for multinational biotech innovators.

The vast majority of South Africa's population has been totally oblivious, uninformed or indifferent concerning the commercial introduction of GM crops and food, although the debate remains highly polarized and adversarial between the pro and anti groups. Activists on both sides indignantly reject the validity of their opponents' arguments, claiming the moral high ground while the government plays a rather passive role (Wolson, 2005a).

The two main anti-GM advocacy groups, Biowatch and The South African Freeze Alliance on Genetic Engineering (SAFeAGE), funded mainly by European NGOs and the German Technical Co-operation Agency (GTZ), oppose GM products on health, environmental and socio-economic grounds. SAFeAGE calls on the government to impose a minimum five-year moratorium on field trials and commercial releases of GM crops in order to ensure that, when released, the technology will have been proven safe, environmentally harmless and in the interest of the South African people. Through a court case, Biowatch has recently been able to increase the amount and accuracy of information the biosafety regulatory authority has to publicly release regarding permit applications. Also, it is said that the extended review period for 'stacked gene' cotton (more than three years) is mainly due to pressure exerted by anti-GM groups upon the regulators and regulatory process, most likely through the Department of Environmental Affairs. Even though delayed release of a technology has financial implications for biotechnology companies – and potentially more importantly, for farmers – it can be argued that these anti-GM groups play a necessary, if mostly unappreciated, watchdog role in keeping the regulatory system and seed and biotechnology companies honest.

South Africa's commercial farmers, hungry for technology that can increase efficiency and profitability, are operating in an environment where production credit and information is available. The major challenge for commercial dissemination of GM crops in the country is how to make these technologies accessible to small-scale farmers, who can greatly benefit, but struggle to afford the technology and inputs that accompany them. As was shown in the Makhatini Flats example, for a crop like cotton, where the harvest can be used as collateral, expensive inputs can be purchased with credit. However, for food crops like maize, no production credits are available to subsistence farmers as these crops are produced for household consumption. Land, in most instances, cannot serve as collateral either because the vast majority of small-scale farmers do not own the land on which they live and farm; it belongs to the local tribe and farmers merely have permission to occupy it.

Results have shown that small-scale farmers benefit from the use of GM maize, in many instances more than large-scale farmers. By employing conservation tillage, controlling weeds with chemical herbicides and using herbicide tolerant varieties, farmers should also be able to decrease the need for expensive, mechanized land cultivation, as well as for labour required for weeding, usually done by family members. With an increasing number of rural households run by elders and children owing to a significant percentage of the economically active family members working in cities, and because the HIV/AIDS infection rate is estimated to be above 16 per cent in the 15–49 age group, farm labour is increasingly expensive and inefficient. Monsanto has organized days for small-scale farmers demonstrating the virtues of conservation tillage and markets input packages with seeds and a corresponding quantity of herbicide, specifically aimed at them. The ARC has also been advocating conservation tillage practices. Government extension officers in some areas have been trained to teach and assist small-scale farmers in using knapsack sprayers and mixing herbicides, and a significant number of maize and cotton farmers in KZN have started buying and planting herbicide tolerant crops.

Adoption of GM maize hybrids could expand if companies were willing to segment the seed market and charge lower prices to small-scale farmers than to large-scale commercial farmers. Pannar currently offers such a programme with some of its conventional hybrid seed: it produces some less expensive seeds for double-cross hybrids or open pollinated varieties and sells them at lower prices to small-scale farmers, at the same time producing high-cost (single-cross) hybrids and selling them at a premium to commercial farmers. However, this pricing strategy may not be possible if the biosafety regulatory requirements are structured so that it becomes more expensive to provide new technology to small-scale farmers than to large-scale producers. Under the current system, every farmer who buys GM seed has to sign a contract with the company selling this seed ensuring that the farmer will plant in the specified location and abide by the stated refuge requirements. This is relatively easy when dealing with large-scale farmers. But if thousands of subsistence farmers have to sign contracts the system could easily become an expensive administrative nightmare, leading to the end of marketing of GM seed to all small-scale farmers.

GM free export markets

South Africa exports maize to countries that demand non-GM maize, including Japan, Zimbabwe, Zambia, Malawi, Mauritius, Kenya, and Mozambique. This requires separating GM from non-GM maize, which South Africa has the infrastructure and know-how to do, along with ample silo facilities owing to the previous government's emphasis on self-sufficiency. For decades the maize sector has been geared to separate white and yellow maize, with yellow produced mainly for animal feed and white predominantly for human consumption.

Where non-GM maize is required, farmers have to declare whether they are delivering GM or non-GM maize, the product is tested using an inexpensive (but arguably rather inaccurate) 'dipstick' test and then delivered to specific silos. There is thus a real effort to segregate GM and non-GM maize, although the system currently depends greatly on the honesty of the farmer and sometimes intended GM tolerance levels cannot be met from certain silos or production areas, especially with increasingly severe tolerance levels from the EU. An increase in a non-GM premium can be expected to increase farmer dishonesty, but at the same time, if premiums increase it might become economically viable to implement a more accurate and stringent segregation system. Contracting farmers to produce non-GM maize has become a common practice among South African food processors and trading companies who have identified a market opportunity to deliver non-GM maize and maize products to domestic consumers or export markets. There are also a number of areas in SA where farmers have decided not to plant GM maize, either as a marketing strategy or just because the technology is not needed in these areas.

Labelling

In 2004, new regulations governing the labelling of foodstuffs were adopted. These regulations cover foods composed of, containing or produced from GM crops, including food additives and ingredients but excluding foods derived from a non-GM animal fed on GM feed. The regulations adopt the US approach of 'substantial equivalent' (Wolson, 2005a). There are currently no GM foods on the market that need to be labelled.

SA does not currently have an official identity preservation system in place, but standards and methods are currently being developed with the assistance of the South African Bureau of Standards (SABS). It has been stressed by the Department of Health that the labelling regulations were only introduced as an interim measure to provide realistic consumer protection pending further investigation (Wolson, 2005a). It is expected that the regulations might be amended based on requirements set by *Codex Alimentarius*, to which SA is a signatory, and the finalization of the identity preservation system.

Wolson (2005a) argues that while civil society groups have campaigned for labelling of GM food, there is not much evidence to suggest that this is an important issue for the public at large. According to a consumer survey conducted by HSRC and PUB in 2004 (HSRC, 2005), more than half of the 7000 respondents indicated that they never read information on food labels, except the brand name. Of the respondents, 21 per cent indicated that they would welcome more labelled information on ingredients and the same percentage also wished to see more health information on labels.

Policy challenges and conclusions

The South African government has recognized the role biotechnology can play in agriculture, food security, rural development and poverty alleviation and has stressed the importance of the creation of a knowledge-based 'bioeconomy' for SA as an international player and leader in Africa. The government faces the considerable challenge of creating a policy, legislative and R&D environment in which large-scale commercial farmers can produce affordable food for the increasingly urbanized South African population and the famine-prone Southern Africa Development Community (SADC) region, where small-scale and emerging, previously disadvantaged farmers can produce agricultural products of value in a sustainable manner, and where subsistence farmers can increase their food security. At the same time SA farmers should be able to produce agricultural products for the international and niche markets where premiums can be earned.

South African agriculture is battling to transform previously disadvantaged farmers and bring them into the mainstream of the agricultural economy. A land redistribution programme is underway and important policy changes are being contemplated to improve the support framework to assist new and smaller scale farmers to become truly independent farm entrepreneurs. A new financing scheme for small-scale farmers is taking shape and might enable them to afford the new technologies that could help them produce in a sustainable, risk minimizing and cost- and labour-efficient manner.

Despite an impressive biotechnology research history, there is currently great concern about the capacity and ability of the ARC to deliver and continue to provide sound and effective research in this field. In the last couple of years a lack of funding, human resources and capacity, as well as poor management, has plagued this organization, affecting the whole agricultural research system. A redesigned institutional and funding framework is urgently needed to ensure that the ARC can continue its vital role as a leading provider of agricultural research in South Africa.

By establishing regional and national innovation centres to facilitate the supply and sharing of capital, expertise, equipment and facilities, as well as support in the commercialization of new products, the government has endeavoured to stimulate product-focused R&D and private investment in biotechnology and agricultural biotechnology R&D. These initiatives aim to create partnerships to share information and alleviate the strain of finding start-up and seed money. The incentives and initiatives created through the National Biotechnology Strategy are certainly positive and a step in the right direction. While the R450 million (US$75 million) made available by the government to implement the Strategy is substantial in African terms, it remains small by international standards (Wolson, 2005) and a significant amount of this money has been and will be spent on establishing the centres. More funds are needed to stimulate development and commercial dissemination.

The government is encouraging public–private research partnerships but joint intellectual ownership and benefit sharing have not yet been clarified. Development of formal legislation is, however, underway. SA has a vast biodiversity that presents immeasurable possibilities and opportunities to biotechnological entrepreneurs. It is therefore extremely important that the country ensure that the correct legislation and institutions are in place to protect and utilize this resource to the benefit of the environment and South Africans. The Biotechnology Partnership and Development and the Bioresource Centres have been tasked with this responsibility.

The Makhatini Flats GM cotton story can serve as a good example for developing countries that are still pondering the possible release of GM cotton. The Makhatini Flats' situation, with its technological triumph but institutional failure, emphasizes the importance of governance, institutional structures and cooperation for the successful adoption and dissemination of GM crops. It also serves as a reminder that, in most cases, achieving scientific advances is really easier than establishing the social and economic conditions necessary for progress.

Notes

1 Coetzee, A., GWK Extension Officer, personal communication (September, 2004).
2 Paterson, A., quoted on www.checkbiotech.org (8 April 2005).
3 Bennett, A., head biotechnologist, Monsanto, South Africa, personal communication (February, 2006).

References

Annecke, D. P. and Moran, V. C. (1982) *Insects and Mites of Cultivated Plants in South Africa*, Butterworths, Durban

Bennett, A. (2002) 'The Impact of *Bt*-Cotton on Small Holder Production in the Makhathini Flats, South Africa', Monsanto, www.monsantoafrica.com

DBSA (Development Bank, South Africa) (1993) 'Statistics on living standards and development: Regional poverty profile', Eastern and Northern Transvaal Policy, Working Paper No 3, DBSA, Midrand

DACST (Department of Arts, Culture, Science and Technology) (1996) 'The White Paper on science and technology', Department of Arts, Culture, Science and Technology, Pretoria

DACST (2001) 'A national biotechnology strategy', Department of Arts, Culture, Science and Technology, Pretoria

DST (Department of Science and Technology) (2004) 'Biotechnology platforms: A strategic review and forecast', Department of Science and Technology, Pretoria

Gouse, M., Kirsten, J. F. and Jenkins, L. (2003) 'Bt cotton in South Africa: Adoption amongst small-scale and large-scale farmers', *Agrekon*, vol 42 (1), pp15–28

Gouse, M. and Kirsten, J. F. (2004) 'Production of genetically modified maize under smallholder conditions in South Africa. Assessment based on three research seasons', unpublished report to Monsanto

Gouse, M., Kirsten, J. F., Shankar, B. and Thirtle, C. (2005a) '*Bt* cotton in Kwa-Zulu Natal: Technological triumph but institutional failure', *AgBiotechNet*, vol 7 (134), pp1–7

Gouse, M., Pray, C., Kirsten, J. F. and Schimmelpfennig, D. (2005b) 'A GM subsistence crop in Africa: The case of *Bt* white maize in South Africa', *International Journal of Biotechnology*, vol 7 (1/2/3), pp84–94

Gouse, M., Pray, E. and Schimmelpfennig, D (2005c) 'The distribution of benefits from *Bt* cotton adoption in South Africa', *AgBioForum*, vol 7 (4), pp1–8, www.agbioform.org

Government of South Africa (2001) *National Biotechnology Strategy for South Africa 2001*, Government of South Africa, Pretoria

Hofs, J., Fok, M. and Gouse, M. (2005) 'Makhatini cotton experience: Updated', forthcoming working paper, URAD and University of Pretoria

HSRC (2005) 'HSRC client survey – 2004', Report to public understanding of biotechnology, available at www.pub.ac.za/resources/docs/

Kirsten, J. F. (2006) 'Socio-economic dynamics of the South African agricultural sector', South African Institute of International Affairs, Trade Policy Brief No 10, available at www.saia.org.za

Liebenberg, G. F., Bientema, N. M. and Kirsten, J. F. (2004) *South Africa, Agricultural Science and Technology Indicators*, Country Brief No 14, International Food Policy Research Institute and the International Service for National Agricultural Research

Meyer, F. H. (2002) 'Modelling the market outlook and policy alternatives for the wheat sector in South Africa', thesis submitted in partial fulfilment of the requirements for the degree MSc, Agriculture, Department of Agricultural Economics, Extension and Rural Development, University of Pretoria, Pretoria

Qaim, M., Cap, E. J. and De Janvry, A. (2003) 'Agronomics and sustainability of transgenic cotton in Argentina', *AgBioForum*, vol 6 (1&2), pp41–47, www.agbioform.org

Shankar, B. and Thirtle, C. (2005) 'Pesticide productivity and transgenic cotton technology: the South African smallholder case', *Journal of Agricultural Economics*, vol 56 (1), pp97–115

Thomson, J. A. (2002) *Genes for Africa: Genetically Modified Crops in the Developing World*, UCT Press, Cape Town

Thomson, J. A. (2004) 'The status of plant biotechnology in Africa', *AgbioForum*, vol 7 (1&2), pp9–12, www.agbioforum.org

Van Der Walt, W. J. and Koster, B. (2005) 'An overview of plant variety protection in South Africa', *IP Strategy Today*, No 13, pp18–27

Wolson, R. A. (2005a) 'Country study South Africa', Report to New York University Project on International GMO Regulatory Conflicts, New York

Wolson, R. A. (2005b) 'Towards establishment of a vibrant South African biotechnology industry: Will the recent policy interventions achieve their objective?', *International Journal of Biotechnology*, vol 7 (1/2/3), pp147–160

Part Three

Comparing and Analysing Developing Country Experiences

Institutional Changes in Argentina, Brazil, China, India and South Africa

Sakiko Fukuda-Parr

Chapter 2 analysed how the emergence and spread of GM crops at the global level was driven by markets and shaped by institutions. This chapter analyses the same process at the national level and compares the experiences of Argentina, Brazil, China, India and South Africa.

Adoption of GM crops requires significant institutional change in two areas: new arrangements for R&D and a regulated seed market for biosafety and IPR enforcement (see Chapter 2). The five countries have similarities but also differences in the challenges they have faced and the approaches they have taken in both these areas, such as with respect to the role of the private and public sectors in research and the way that patents and biosafety rules are implemented. These institutional innovations not only differ from one another but from the US model. This chapter first reviews the evolution of commercial production and the economic factors that have driven it; then it looks at the institutional evolution in R&D and in seed marketing. The last section explores the institutional options for developing countries, contrasting the two dominant approaches that have emerged, namely the corporate and the 'super NARS' business models.

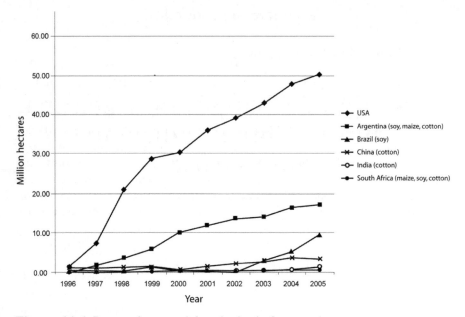

Figure 11.1 *Pattern of commercial production in five countries*
Source: James, 2005

Table 11.1 *Adoption rates: Percentage of total acreage sown to GM varieties, 2004*

	RR Soy	*Bt* and RR Cotton	RR and *Bt* Maize
US	85	76	45
Argentina	98	20–25	55
Brazil	22	0	0
China	0	66	0
India	0	6	0
South Africa	35	81 (*Bt*) 7 (RR)	27 (*Bt* yellow maize) 8 (*Bt* white maize)

Source: South Africa from Gouse (Chapter 10); Argentina, Brazil, China and India from James (2005); Cotton estimates for China and India based on projections 2002 *Bulletin of the International Cotton Advisory Committee ICAC* (www.icac.org); US from 'Pew Initiative on Food and Biotechnology (pewagbiotech.org) Fact sheet: Genetically Modified Crops in the United States', August 2004. Other sources give divergent figures: Brazil RR soy 42 per cent if calculated from FAO total crop acreage, China *Bt* cotton 69 per cent from ICAC, and India 15 per cent from ICAC

Commercial production

Trends: Diffusion outpaces approvals

Figure 11.1 shows the pattern of commercial production in the five countries while Table 11.1 shows the adoption rates. These numbers may be underestimated in some situations because the varieties have diffused through the informal market for seeds in 'white bags' rather than through the formal market of licensed sales in Argentina, Brazil, China and India. Expansion of total areas has been more rapid for RR soy (Argentina and Brazil) than for *Bt* cotton and maize. In terms of adoption rates, expansion of commercial production has reached over 50 per cent of the area planted for RR soy in Argentina, *Bt* cotton in China and *Bt* cotton in South Africa. But RR soy in Brazil has reached only 22 per cent, *Bt* cotton in Argentina 20–25 per cent, and maize in South Africa 8–27 per cent. The reasons for the lower adoption rates are either a slow start due to a lengthy process of establishing biosafety legislation (cotton in India and soy in Brazil), or low incentives to farmers because the crops do not perform well or do not offer consistently significant advantages in income (maize in South Africa, cotton and maize in Argentina). These challenges of forging national consensus and developing research capacity that is responsive to farmer demands will be explored in the later section and in Chapter 12.

Monsanto negotiated with local seed companies to set up joint ventures and introduce its GM soy and GM cotton to all five countries at about the same time that it released its seeds in the US. By then Chinese NARS had developed *Bt* cotton. In Argentina, Brazil, China and South Africa, commercial production expanded rapidly starting in 1996. But in India and Brazil, commercial releases were held up by mounting opposition and controversies over biosafety legislation. In Brazil, after an initial approval of a RR soy variety in 1996, commercial planting of all transgenics was effectively banned until 2005, when new biosafety legislation was passed. In the interim, however, faced with the anomalous situation in 2003–2004, the government issued temporary authorization for that season.

The contrast between these two sets of countries mirrors the contrast between the US and the EU and led to similar differences in types of biosafety legislation – Argentina, China and South Africa have more permissive legislation, while the legislation in Brazil and in India applies precautionary principles. Why are they so different? The different makeup of interest groups in these countries may be an explanation and will be explored in the next chapter.

However, Brazil and India did not follow the same path as Europe; commercial production was arrested and R&D slowed down in Europe, but in Brazil and India research programmes continued, informal sector seed supplies developed and farmers started growing these varieties irrespective of legislation. In Brazil, the seeds were brought in from neighbouring Argentina, and in India, they were developed and sold by entrepreneurial seed companies.

Within country differences in diffusion rates

Diffusion rates of GM crops that have been introduced have varied within countries – from one crop to another, and from region to region. These differences can be explained by varying levels of farmer demand.

In Brazil, the diffusion of RR soy has been concentrated in the south (states of Rio Grande do Sul and Paraná) and limited in the centre west (states of Mato Grosso and Goiás). As the case study shows (see Chapter 7), this is because the RR soy variety, imported from Argentina, is not well adapted to growing conditions in the centre west, resulting in few benefits for farmers. Da Silveira and Carvalho Borges speculate that varieties that are better suited to the centre west and to Brazilian conditions in general will be developed and released now that the legislative framework has been approved. GM maize and cotton have not spread in Brazil in the same way that RR soy has. Why not? Perhaps it is because an informal supply of seeds has not been available, or perhaps because farmer demand has not been strong.

In South Africa, *Bt* maize occupies only 27 per cent of the yellow maize and 8 per cent of the white maize areas, which contrasts with the more rapid spread of *Bt* cotton, which occupied over 80 per cent of cotton areas in 2003/2004. At first, the *Bt* yellow maize that was supplied was the US variety that had not been adapted. Local varieties were developed, but adoption rates levelled off with farmers recognizing that net income benefits of *Bt* maize depended on the level of stalk borer infestation.

In Argentina, the rapid diffusion of RR soy contrasts with the slow diffusion of *Bt* cotton and *Bt* maize. Chudnovsky (see Chapter 6) argues that an important factor in the diffusion of RR soy was the absence of patent protection, which put downward pressure on seed prices. As he explains, Nidera is able to market RR soy without a licence. There has also been a proliferation of 'white bag' supplies for a crop that reproduces itself. Furthermore, the Argentine intellectual property framework protects farmers' right to keep seeds for their own use, following the UPOV convention. As soy is self propagating, both farmers and seed entrepreneurs can easily reproduce GM varieties. This contrasts with the situation for maize and cotton, for which Monsanto patents are rigorously enforced, and for crops that are difficult to multiply. While soy is autogamous, *Bt* maize is a hybrid and cotton seeds need special treatment. Thus, it would be difficult for farmers to keep their own seeds, though these technical obstacles would not stop an entrepreneurial seed company with technological know-how from multiplying *Bt* maize and *Bt* cotton seeds off patent, as has been done in India.

Raney (2006) also concludes that patents and the high cost of seeds were critical factors in the slow spread of *Bt* cotton and the rapid spread of RR soy in Argentina. Table 11.2 shows the advantage of *Bt* cotton over conventional varieties in Argentina, China, India, Mexico and South Africa, drawing on evidence from ex post farm impact studies published in peer reviewed literature to date. Seed costs in Argentina for *Bt* cotton are five times the costs for

Table 11.2 *Performance advantage of Bt cotton over conventional varieties (%)*

	Argentina	China	India	Mexico	South Africa
Yield	33	19	34	11	65
Revenue	34	23	33	9	65
Pesticide costs	−47	−67	−41	−77	−58
Seed costs	530	95	17	165	89
Profit	31	340	69	12	299

Source: Raney (2006)

conventional varieties, while in China they are 95 per cent more expensive and in India only 17 per cent more.

In South Africa, while *Bt* cotton has spread rapidly, this spread had different patterns in the two sectors of the country's dual agricultural structure. Keen to show the relevance of the technology to the traditional sector, Monsanto introduced *Bt* cotton varieties to the Makhatini Flats small-scale farming communities in 1997/1998. Adoption increased rapidly in the following two years, with farmers enjoying substantial yield increases that were more than those realized by larger scale commercial farmers. But these gains were not sustained, not because the varieties do not perform well agronomically, but because farmers did not obtain credit when their marketing and credit agreements collapsed.

Institutional factors in diffusion

These experiences highlight the important role of some institutional factors in the diffusion of GM crops. Farmers are best served by a competitive market that can supply seeds of varieties that perform well under local conditions, are offered at low prices and are supported by input supply and marketing services. The ability of local seed companies and research establishments to develop a stream of locally adapted varieties, using local cultivars, plays a key role in providing farmers with seeds that meet their needs. Depending on imported varieties – whether Argentinean seeds in Brazil or US seeds in South Africa – has its limits for supplying farmers with seeds of good quality and variety that perform well in the local conditions and address constraints that they face. The informal sector – whether the entrepreneurial seed companies or farmers themselves – has created a competitive supply with low prices for farmers. But these informal sector seeds, too, are not ideal for meeting farmer needs; 'white bags' cannot communicate reliably what is being sold, as Ramaswami and Pray point out, the quality is likely to be compromised, and their proliferation may wipe out the formal sector supply because they are priced lower and do not include the licensing and other technology fees.

The experience of China illustrates these points. *Bt* cotton is supported by an extensive network of public research institutions throughout the country, each working on developing locally adapted varieties. They have released hundreds of varieties of *Bt* cotton. The first among them were approved in 1997, of which four were developed by the Chinese public research system and one by Monsanto in a joint venture with a local seed company. Seed prices are kept low, much lower in China than elsewhere (Raney, 2006). There is no technology fee to collect because R&D is publicly financed. For these reasons, gains for farmers are higher in China than in Argentina, India, Mexico and South Africa (see Table 11.2). The case of Makhitini Flats in South Africa shows the particular constraints of subsistence farmers who are not well served by credit and marketing systems.

Developing R&D capacity

The R&D capacity of developing countries to access modern biotechnology has long been in question when considering whether this technology could be useful for developing countries (Byerlee and Fisher, 2001; UNDP, 2001). Admittedly, China, Brazil and India have the largest and the most dynamic research systems in the developing world. Nonetheless, the experience of the five countries indicates that the obstacles to developing country access may not be as overwhelming as has been anticipated. Developing biotechnology capacity may be difficult but accessing products developed elsewhere and developing locally adapted varieties has been scientifically within reach. All five countries are investing in national R&D capacity, but their outputs vary in terms of commercial viability of products, scope of investments and institutional approaches followed.

Outputs

Complete data on commercial varieties that have been developed and approved are difficult to compile. Nevertheless, Table 11.3 is an attempt in this direction, drawing on published sources and information from the case studies and their authors. Though it may not be fully complete, it gives a rough approximation of the basic trend and key features.[1]

First, all of the approvals in all countries are for multinationals and their joint ventures with local companies, except in China, where the public sector dominates, and in the US, where universities and the USDA have developed commercial varieties. But the latter are for minor crops – papaya and flax – grown over very small areas.

Second, China is so far the only one of the five countries to have developed GM varieties that are in commercial production, starting from the biotechnology step. It has a decade of experience, having released its first varieties in

Table 11.3 *Varieties approved for commercialization and associated transformation events*

Country	Crop	Year approved for commer- cialization	Event	Trait	Seed company
Argentina	Soy	1996	40-3-2	Herbicide tolerance	Nidera SA (Monsanto)
	Maize	1998	176	Insect resistance	Ciba-Geigy (Syngenta)
	Maize	1998	T25	Herbicide tolerance	AgrEvo SA (Bayer, Aventis)
	Cotton	1998	MON 531	Insect resistance	Monsanto Argentina SAIC
	Maize	1998	MON 810	Insect resistance	Monsanto Argentina SAIC
	Cotton	2001	MON 1445	Herbicide tolerance	Monsanto Argentina SAIC
	Maize	2001	*Bt*11	Insect resistance	Novartis Agrosem SA (Syngenta)
	Maize	2004	NK603	Herbicide tolerance	Monsanto Argentina SAIC
	Maize	2005	TC1507	Stacked: insect resistance, herbicide tolerance	Dow Agrosciences Argentina SA Pioneer Argentina SA
	Maize	2005	GA21	Herbicide tolerance	Syngenta
Brazil	Soy	1996*	40-3-2	Herbicide tolerance	Monsanto
	Cotton	2005	MON 531	Insect resistance	Monsanto

*Subsequently, legal decisions imposed restriction on growing of all GM varieties until 2005

China	Cotton	1997		Insect resistance	Monsanto (two joint venture seed companies with Jidai and Andai)
	Cotton	1997		Insect resistance	4 varieties CAAS
	Cotton*	1998 to present	*numerous	Insect resistance	Mostly CAAS
	Tomato, sweet pepper and petunia*	1998 to present	*numerous		Mostly CAAS

*China approved 40 events of *Bt* cotton for *Bt* cotton commercialization and about ten or more events for tomato, sweet pepper and petunia in 1997–2003. In 2004, there were about 130 events approved (in that one year), most *Bt* cotton, and most from CAAS and a few from Monsanto. Total number of varieties approved for commercialization reached about 50 by 2004.

India	Cotton	2002	MON 531	Insect resistance *Bt*	Mahyco Monsanto Biotech
	Cotton	2004	MON 531	*Bt*	Raasi Seeds (Monsanto jv)
	Cotton	2005	MON 531	*Bt*	Rasi Seeds (Monsanto jv)
	Cotton	2005	MON 531	*Bt*	Mahyco Monsanto Biotech
	Cotton	2005	MON 531		Ankur Seeds
	Cotton	2005	MON 531	*Bt*	Nuziveedu Seeds
	Cotton	2005	MON 531	*Bt*	Mahyco Monsanto Biotech
	Cotton	2005	MON 531	*Bt*	Nuziveedu Seeds
	Cotton	2006	MON 531	*Bt*	Ganga Kaveri
	Cotton	2006	MON 531	*Bt*	Ajeet Seeds

Cotton	2006	MON 15985	*Bt*		Ajeet Seeds
Cotton	2006	MON 531	*Bt*		Rasi Seeds
Cotton	2006	MON 531	*Bt*		Emergent Seeds
Cotton	2006	MON 531	*Bt*		Nuziveedu Seeds
Cotton	2006	MON 531	*Bt*		Pravardhan Seeds
Cotton	2006	MON 531	*Bt*		Prabhat Seeds
Cotton	2006	MON 531	*Bt*		Krishidhan Seeds
Cotton	2006	MON 15985	*Bt*		Krishidhan Seeds
Cotton	2006	MON 15985	*Bt*		Mahyco Monsanto Biotech
Cotton	2006	(Cry1Ab + CryAc) GFM	*Bt* and RR		Nath Seeds
Cotton	2006	Event 1 (Cry1Ac)	*Bt*		JK Seeds
Cotton	2006		*Bt*		Numerous*

*June 2006, 59 *Bt* hybrids were approved of which 52 use Monsanto genes under license; others use genes from China and India.

South Africa	Cotton	1997	MON 531/ 757/ 1076	Insect resistance	Monsanto (Delta & Pine Land)
	Maize	1997 (white and yellow)	MON 810	Insect resistance	Monsanto
	Soy	2001	40-3-2	Herbicide tolerance	Monsanto
	Cotton	2001	MON 1445	Herbicide tolerance	Monsanto (Delta & Pine Land)
	Maize	2003	*Bt* 11	Insect resistance	Novartis Agrosem S.A. (Syngenta)
	Maize	2003	NK 603	Herbicide tolerance	Monsanto

	Crop	Year	Line	Trait	Company
	Cotton	2005		*Bt*, RR stacked	Monsanto (Delta & Pine Land)
EU	Maize	1997	*Bt*-176	Insect resistance	Ciba-Geigy (Syngenta)
	Canola	1997	MS1, RF2	Stacked: herbicide tolerant and pollination control	Aventis
	Maize	1998	T25	Herbicide tolerance	AgrEvo
		1998	MON 810	Insect resistance	Monsanto
	Maize	2004	*Bt*11	Insect resistance	Syngenta
	Maize	2006	DAS1507	Stacked; insect resistant and herbicide tolerant	Mycogen (Dow, Pioneer)
US	Tomato	1992	FLAVR SAVR	Fruit ripening altered	Calgene
	Squash	1992	ZW-20	Virus resistant	Upjohn
	Cotton	1993	BXN	Herbicide tolerant	Calgene
	Soybean	1993	40-3-2	Herbicide tolerant	Monsanto
	Canola	1994	pCGN 3828-212/86-18 & 23	Oil profile altered	Calgene
	Tomato	1994	Line N73 1436-111	Fruit ripening altered	Calgene
	Tomato	1994	1345-4	Fruit ripening altered	DNA Plant Tech
	Tomato	1994	9 additional FLAVRSAVR lines	Fruit ripening altered	Calgene

Potato	1994	*Bt*6, *Bt*10, *Bt*12, *Bt*16, *Bt*17, *Bt*18, *Bt*23	Insect resistant	Monsanto
Tomato	1994	B, Da, F	Fruit ripening altered	Zeneca & Petoseed
Cotton	1994	531, 757, 1076	Insect resistant	Monsanto
Maize	1994	Event 176	Insect resistant	Ciba Seeds
Maize	1994	T14, T25	Herbicide tolerant	AgrEvo
Tomato	1995	20 additional FLAVRSAVR lines	Fruit ripening altered	Calgene
Cotton	1995	1445, 1698	Herbicide tolerant	Monsanto
Tomato	1995	8338	Fruit ripening altered	Monsanto
Maize	1995	MON 80100	Insect resistant	Monsanto
Maize	1995	B16	Herbicide tolerant	DeKalb
Tomato	1995	2 additional FLAVRSAVR lines	Fruit ripening altered	Calgene
Maize	1995	*Bt*11	Insect resistant	Northrup King
Maize	1995	MS3	Male sterile	Plant Genetic Systems
Cotton	1995	19-51a	Herbicide tolerant	Du Pont
Tomato	1995	35 1 N	Fruit ripening altered	Agritope
Potato	1995	SBT02-5 & -7, ATBT04-6 &-27, -30, -31, -36	Insect resistant	Monsanto
Squash	1995	CZW-3	Virus resistant	Asgrow
Maize	1996	MON809 & MON810	Insect resistant	Monsanto
Papaya	1996	55-1, 63-1	Virus resistant	Cornell U

Soybean	1996	W62, W98, A2704- 12, A2704-21, A5547-35	Herbicide tolerant	AgrEvo
Tomato	1996	1 additional FLAVRSAVR line	Fruit ripening altered	Calgene
Maize	1996	DBT418	Insect resistant	DeKalb
Maize	1996	MON802	Stacked: herbicide tolerant and insect resistant	Monsanto
Soybean	1997	G94-1, G94-19, G-168	Oil profile altered	Du Pont
Cotton	1997	Events 31807 & 31808	Stacked herbicide tolerant and insect resistant	Calgene
Maize	1997	GA21	Herbicide tolerant	Monsanto
Cichorium intybus	1997	RM3-3, RM3-4, RM3-6	Male sterile	Bejo
Potato	1997	RBMT21-129 & RBMT21-350	Stacked: insect and virus resistant	Monsanto
Canola	1997	T45	Herbicide tolerant	AgrEvo
Maize	1997	CBH-351	Stacked: herbicide tolerant and insect resistant	AgrEvo
Tomato	1997	5345	Insect resistant	Monsanto
Beet	1997	T-120-7	Herbicide tolerant	AgrEvo
Potato	1997	RBMT15-101, SEMT15-02, SEMT15-15	Stacked: insect resistant, virus resistant	Monsanto
Maize	1997	676, 678, 680	Stacked: Male sterile and herbicide tolerant	Pioneer
Soybean	1998	A5547-127	Herbicide tolerant	AgrEvo

Beet	1998	GTSB77	Herbicide tolerant	Novartis Seeds & Monsanto
Canola	1998	RT73	Herbicide tolerant	Monsanto
Soybean	1998	GU262	Herbicide tolerant	AgrEvo
Canola	1998	MS8 & RF3	Stacked: herbicide tolerant and Pollination control	AgrEvo
Rice	1998	LLRICE06, LLRICE62	Herbicide tolerant	AgrEvo
Flax	1998	CDC Triffid	Tolerant to soil residues of sulfonyl urea herbicide	U. of Saskatchewan
Maize	1998	MS6	Staked: herbicide tolerant and Male sterile	AgrEvo
Potato	1999	RBMT22-82	Stacked: virus and insect resistant	Monsanto
Maize	2000	NK603	Herbicide tolerant	Monsanto
Maize	2000	Line 1507	Stacked; insect resistant and herbicide tolerant	Mycogen c/o Dow & Pioneer
Cotton	2000	Cotton Event 15985	Insect resistant	Monsanto
Tobacco	2001	Vector 21-41	Reduced nicotine	Vector
Maize	2001	MON 863	Insect Resistant	Monsanto
Canola	2001	MS1 & RF1/ RF2	Stacked: herbicide tolerant and pollination control	Aventis
Canola	2001	Topas 19/2	Herbicide tolerant	Aventis
Canola	2001	RT200	Herbicide tolerant	Monsanto

Cotton	2002	LLCotton25	Herbicide tolerant	Aventis
Cotton	2003	281-24-236	Insect Resistant	Mycogen/Dow
Cotton	2003	3006-210-23	Insect Resistant	Mycogen/Dow
Cotton	2003	COT 102	Insect Resistant	Syngenta
Maize	2003	TC-6275	Stacked: Insect Resistant herbicide Tolerant	Dow
Sugar beet	2003	H7-1	Glyphosate Tolerant	Monsanto
Maize	2003	59122	Maize Rootworm Resistant	Dow
Cotton	2004	MON 88913	Herbicide Tolerant	Monsanto
Alfalfa	2004	J101, J163	Herbicide Tolerant	Monsanto & Forage Genetics
Maize	2004	88017	Insect Resistant	Monsanto
Maize	2004	LY038	High Lysine	Monsanto

Note: jv = joint venture.
Source: AgBios Biotech Crop Database, updated from personal correspondence with
J. Huang for China and B. Rhamaswami for India

1997 at the same time as a Monsanto variety. The outputs of the Chinese R&D system rival, if not surpass, those of the US in sheer quantity. Huang et al (see Chapter 8) report approvals in the hundreds, with 130 events approved in just one year, 2004, while in the US (see Chapter 3) over 500 varieties have been released. By contrast, so far the Chinese releases are all in *Bt* cotton, with the exception of petunia.

But the situation is rapidly changing with the legislative changes in Brazil and India. As this volume goes to press, India has just approved over 59 new varieties of *Bt* cotton developed by Indian seed companies. What is important is that while 52 of them licensed the gene from Monsanto, the others were developed with genes from Chinese and Indian institutions.[2] Evidently, India is rapidly catching up and becoming another competitor to Monsanto, Syngenta and CAAS. Brazil, by contrast, has developed a private–public partnership in R&D to develop new varieties.[3]

Other crops are in advanced stages of field trials and approval in China, Brazil and India. Rice varieties have been submitted for approval but withdrawn. A new generation of GM crops is likely to appear from Brazil and India, which already have varieties in the field trial stage.

Scope of investment

The size of the Chinese R&D enterprise in GM technology is substantial, comparable to that of the multinationals. Investments of US$200 million (or US$953 million in PPP US$) compare with Monsanto's US$588 million. But what is particularly striking is the extensiveness of the network, involving not just a central research body but provincial ministries and research institutions and stations, with 150 laboratories involved by 2001. The number of personnel concerned is in the thousands, totalling 3200 professionals by 2003. Though not directly comparable, the US case study reports about 1600 crop improvement scientists employed in the US in both private and public sectors in 2001, of which 22 per cent were engaged in biotechnology-related research. These numbers show investment in China that is competitive to that in the US, as do the number of varieties that have been made available to farmers.

Just as Monsanto reinvented itself as a biotechnology seeds company, so did the Chinese Academy of Agricultural Sciences and its network of NARS. Its programmes and capacity grew rapidly starting from the late 1980s, with investments in this area climbing from US$33 million in 1995 to US$200 million by 2003, and further since. While Monsanto and other private sector industries responded to market incentives created by the advances of science, it was deliberate public policy that was behind China's technology development. China has a well-articulated policy towards agricultural biotechnology as a priority, focused on its most economically and socially important crops: cotton and rice, the most important cash and food crops. The policy is also well articulated as a part of the country's overall policy goal: developing China as a leading country in science and technology that is competitive in global markets. As Huang et al state, 'the goal is to make China one of the world's premiere centres for plant biotechnology research – an ambition embodied in the country's 11th Five Year Plan 2006–2010' (Chapter 7).

Brazil and India have similar policy goals for leading global science and technology, particularly in agriculture, and have a long history of agricultural research. They clearly have ready plant breeding capacity to take advantage of the spillover benefits of global advances in GM technology. However, they have been slower to develop their biotechnology R&D programmes and the extent of their effort and financial commitments is not comparable to China's. Brazil's total budget for EMBRAPA is in the order of US$300 million per year, of which about US$30 million is estimated to be for agricultural biotechnology. Moreover, their policies are much less well articulated in terms of thematic focus. Another major constraint to the release of commercial products has been the absence of a clear biosafety framework and the underlying difficulties in achieving national consensus over the issue. Ramaswami and Pray refer to official support in India as 'sporadic' (Chapter 8).

South Africa has adopted a proactive policy towards science and technology as a pillar of development, but its R&D in agricultural biotechnology has been limited in comparison to that of the other countries. Not only are the levels of investment modest – with a total of US$75 million budgeted for four years for all biotechnology (see Chapter 10) – the focus has been on upstream biotechnology research rather than on developing commercially viable crops or tools, as in China, Brazil and India. Neither has South Africa articulated clear roles for its public sector vis à vis the private multinationals and the national private sector.

Of the five countries, Argentina has the smallest R&D programme with little if any involvement of the public agricultural research institutions in the development of existing or new releases of GM varieties. The research that goes on is focused on the plant breeding step and done by seed companies. Despite the official commitment to strengthening agricultural research in both the public or the private sectors effective support is very limited.

Institutional approaches

There are institutional approaches through which countries can access and benefit from agricultural biotechnology: relying more or less on their own national capacities or drawing from global R&D, and relying more or less on public institutions or on private investment. The scientific process of developing a commercially viable GM variety involves two distinct steps (Box 2.1): a biotechnology step leading to a successful 'transformation event', and a plant breeding step leading to a variety adapted to a particular location. These two steps do not need to be carried out in the same country and a country can take advantage of the spillover benefits of the biotechnology stage R&D by licensing the technology. The US and EU countries have developed capacity in both processes in their corporate sectors, and in addition, developed upstream biotechnology research in their academic and research sectors. However, it is possible for countries to take advantage of the spillover benefits without developing the biotechnology capacity. The plant breeding step can also be done in the NARS or in the private or public seed companies. Thus countries can develop their in-country capacity in both biotechnology and plant breeding steps, and in the public sector (NARS) or the private sector (seed companies) (Byerlee and Fisher, 2001). This process of developing a commercial product needs to be fed by upstream biotechnology research that can be done in universities and other research institutions or in the private sector, in the biotech start-up companies, or in multinational life science corporations.

While the biotechnology step is scientifically more challenging than the plant breeding step in GM crop development, the plant breeding step is more challenging organizationally for developing countries. The latter results in a variety to be commercialized that requires biosafety approval. For a company to obtain that approval as well as to negotiate a licence from the gene owner is particularly complex. Traxler speculates that in the US, the reason why large multinationals dominate the market is that they have the financial and other resources to push through the regulatory and IPR processes that smaller companies do not.

What have been the sources of R&D for the products that have been commercialized to date?[4] The five countries differ in approach from the US but also from one another, as summarized in Table 11.4. While the US model is characterized by vertical integration in the corporate sector, and a distinct public sector role in upstream biotechnology research, the emerging Chinese model is characterized by a distinct private sector role in seed multiplication and marketing, the rest being done by the public sector but without the vertical integration. The other countries fall in between, but India and Brazil contrast with South Africa and Argentina. India and Brazil, like China, are characterized by a strong public sector role in all phases of product development and upstream research. South Africa and Argentina have taken a more 'leave it to the private sector' approach, taking advantage of the spillover benefits of global R&D without developing their own products from the biotechnology stage.

Developing countries have successfully adopted GM crop varieties by relying on private sector innovation in the seed sector without necessarily developing their own biotechnology capacity in NARS. Argentina and South Africa are relying on this approach – the private sector and global science – to access the benefits of agricultural biotechnology. China has developed both biotechnology and plant breeding capacities in its public sector. Brazil and India are starting later, with fewer resources than China, to develop both biotechnology and plant breeding capacities in the public sector, and plant breeding capacity in the seed companies. India has released a variety using its own event. Brazil has released varieties based on joint venture. These countries are likely to emerge as important sources of new second generation GM crop varieties.

What is noticeable is the absence of institutional arrangements that have been important for creating technological solutions for development:

Table 11.4 *Organization of research, development and commercialization*

	Public/non-profit (research institutes, universities)	Private (multinational corporations, biotech start ups, local seed companies)	Public–private Partnerships
Upstream biotechnology research	US Argentina Brazil China India South Africa	US (MNC, BSU)	South Africa
Biotechnology step in GM crop development – to 'event'	Brazil China India	US (MNC)	Brazil
Plant breeding step in GM crop development – to GM variety	China India	US (MNC) Argentina (Local) China (MNC/JV) Brazil (JV/Local) South Africa	Brazil
Seed multiplication and marketing	China (informal sector)	US (MNC) China (Local) Brazil (Local) India (JV/local)	

private–public partnerships that are playing a central role in the pharmaceuticals sector, and an international public sector that has been an essential source of agricultural technology in the past. Brazil, where EMBRAPA is cooperating with the multinationals, is farthest in pursing private–public partnership.[5] Collaborative efforts are also underway in South Africa with foundations (Gates) and universities but have not yet resulted in concrete products. None of the inputs to the commercial crops have come from the CGIAR institutions. There has also been no diffusion from one developing country to another except through the informal seed markets from Argentina to Brazil.

Creating a regulated seed market: Biosafety controls, intellectual property and seed marketing

By far the most difficult institutional challenge in developing countries has been to develop a regulated seed market in which patents and biosafety regulations are enforced. All five countries have been introducing new and stronger intellectual property legislation in compliance with TRIPS. The countries have also developed new biosafety regulations, procedures and structures for approval of events, field testing and commercial release. In all these countries, these processes have been evolving as the countries gain experience with the introduction of GM crops. In India and Brazil, the process of developing biosafety legislation has been highly contested. Both countries have been – and continue to be – politically gridlocked in a similar way to France and other EU countries, resulting in lengthy and unpredictable processes of approvals. In Brazil, the cleavage is between the environment and agriculture departments of government, with environmental groups contesting the authority of the regulatory structures in which the Ministry of Science and Technology has main responsibility. These policy processes are explored in Chapter 12.

This chapter focuses on the challenge of implementation and enforcement in the context of a regulated seed market, which has been by far the most difficult challenge. All countries except South Africa have experienced a rampant proliferation of seeds from suppliers who have not licensed the products from patent holders and whose products are not certified as the varieties that have been officially released.

Developing a regulated seed market for GM seeds requires a major institutional shift in developing countries where commercial and informal seed markets exist side by side. In most countries, hybrids are supplied by organized commercial seed companies (alongside many small companies) and by farmers themselves. These seed supply systems are intrinsically connected to plant improvement and breeding activities, which take place primarily in NARS and to a more limited extent in seed companies, as well as by farmers themselves, who select and multiply seeds for their own and neighbours' use.[6] In the US, as in the commercial sectors of many developing countries, farmers have a more consistent tradition of buying seed from seed companies. These companies require farmers to sign a contract agreeing not to save their own seeds so they have to buy GM seeds annually from a licensed supplier who supplies only those varieties that have been approved for commercialization. Such agreements are difficult to enforce. In the US, enforcement is handled by the seed companies themselves, which have the incentive to take care of it. In other countries, it is even more difficult to enforce such systems. There is little incentive for farmers not to keep their seeds in the case of soy, and easy for entrepreneurs to set up seed companies that sell GM seeds of cotton and maize, as has happened extensively in India (see Chapter 9; Herring, in press; Scoones 2006).

It is not surprising that the informal sector has proliferated, though in different ways in the four countries. In Brazil, seeds were imported in defiance of a ban in response to strong farmer demand. In Argentina, 'white bags' supply about a third of the market for soy, but have not appeared for maize and cotton, which are difficult to reproduce and for which patents are strictly enforced. In India, private entrepreneurship produced *Bt* cotton seeds. In China, entrepreneurial seed companies and state research stations sold *Bt* cotton seeds to farmers. Why did the informal sector not develop in South Africa? One reason could be that maize and cotton, unlike soy, are not easily multiplied. Another is that cotton seeds were provided as part of a system where a single company purchased the seed cotton and provided seeds and other inputs along with credit.

The US model of a regulated seed market with strict legal enforcement does not fit the realities of developing countries. In India, when *Bt* cotton varieties were found growing before they were authorized, courts ordered them to be burned. But this was not a politically enforceable verdict.

Farmer demand for GM varieties drives the incentives for the informal market in GM seeds. Opposition movements demand precautionary biosafety frameworks that restrict or ban authorized seeds. Ironically, this opens the space for the informal market supply. Farmers lose because they get seeds that may be less effective than those that more investment in R&D could supply. Recent studies in both India and China showed that the informal market seeds did not perform as well as the authorized seeds, though still better than the non-GM conventional varieties (Morse et al, 2005).

Restrictive biosafety frameworks have led to unexpected and perverse results in another way. Comparing the experience of India and China, Pray et al (2006) found obtaining approval much more costly, lengthy and unpredictable in India than in China. As the case study in this volume argues, the result in India is that those varieties that the official market in GM seeds offers are less competitive because only one variety was approved between 2002 and 2006. China encourages seed companies selling unauthorized varieties to go through the approval process. This approach effectively gives incentives for all seed entrepreneurs to integrate into the regulated market rather than to stay out in the informal sector. The result is a system more capable of providing farmers with a greater choice of varieties.

Emerging business models for seed development and commercialization

Chapters 2 and 3 elaborated on the institutional shifts that took place in the US which underpinned the emergence of GM crops:

- the restructuring of a seed delivery system and market dominated by vertically integrated multinationals incorporating the upstream biotechnology and downstream plant breeding phases of R&D to develop a commercial product;
- the development of stronger patent protection for plants and genes;
- the evolution of supportive biosafety legislation with a reasonable level of national consensus and enforcement;
- a supportive government policy including financing of upstream biotechnology research.

Institutional shifts being made in the five developing countries differ from this US model and also from one another. The US model can be characterized as a 'corporate business model' where market incentives have driven corporations to invest heavily in biotechnology research, seed development and marketing. In China, Brazil and India, the public sector is investing in biotechnology research, while a combination of private seed companies and public sector research stations are involved in seed development and marketing. In South Africa and Argentina, investments are focused on seed development and marketing in private seed companies. Investment in biotechnology research is upstream rather than in development of commercial varieties.

In all five countries, new institutional arrangements were made in the seed sector to deliver the new varieties, either with joint ventures between multinationals and local seed companies or public–private partnership with a national research institution. Most importantly, national entrepreneurship also began to import and directly market a variety developed for other markets (from Argentina to Brazil), or to backcross and develop adapted varieties, without official approval, as in India before 2002. Thus neither financial, technological nor organizational weaknesses were constraints to these countries' accessing spillover benefits of global GM crop development. In fact, as Traxler argues, developing locally adapted varieties of GM crops is even easier and less costly than developing conventional varieties, which takes a great deal of time (5–10 years) of cross breeding (see Chapter 3).

By contrast, all the five chapters, except that on China, emphasize financial and scientific resource constraints as significant obstacles to developing full capacity to make use of biotechnology. In the US, this capacity was developed by market incentives that mobilized financial resources to invest in R&D capacity, acquire seed companies and restructure the research/seed sector into a vertically integrated industry. In China, well-articulated government policy on biotechnology as a national priority has effectively mobilized increasing resources to create capacity in a similar way. The Chinese model offers an alternative 'public sector' business model for GM crop R&D and marketing. This model differs from the 'corporate model' in a number of other ways. First, the development priorities do not need to be driven by profitability but by national public priorities. Second, the source of financing is not predicated on collecting a 'technology fee' under patent protection to recoup the investments made in R&D. Thus patenting, licensing and seed pricing can be structured along a different logic. As already noted, public sector *Bt* cotton seeds in China are sold at a fraction of the cost of Monsanto *Bt* cotton seeds in Argentina. This may also explain the extraordinary proliferation of different events and varieties being developed in China, which exceeds the numbers being achieved in the US.

Traxler argues that developing country seed markets are not large enough to offer financial incentives for investment in developing GM varieties (see Chapter 3). But market incentives would not drive public investment priorities. In China, Brazil and India, R&D priorities in biotechnology emphasize crops that are of national importance. Cotton in China is the most important crop, along with rice; cotton in India is also a key cash crop while soy is the most important export crop in Brazil. Public research programmes in Brazil emphasize other high priority crops for both export and domestic consumption, including cassava and beans, which do not attract private investment. So the constraint to financing R&D is the availability of public budgets, and whether these investments can command priority relative to competing needs. Large countries such as China can mobilize resources, but this will be difficult in smaller countries. In the US and EU, where there are industry and farm lobbies that have political power, institutional changes have been made. In developing countries, where farmers do not command much political power, such public mobilization may be difficult.

Notes

1 The table has also been reviewed by researchers knowledgeable about the situation in each of the countries.
2 Rhamaswami, personal communication, June 2006.
3 Da Silveira, personal communication, May 2006.
4 This analysis focuses on products commercialized to date, not on activities that are ongoing but that have not led to a major commercialized product.
5 Other private–public partnerships are noted, such as in South Africa. However, these have not led to major commercial products.
6 See Louwaars et al (2005) for comprehensive review of biotechnology and seed systems, including existing systems and the requirements of GM crops.

References

AgBios, Biotech Database, www.agbios.com accessed May 2006
Byerlee, D. and Fischer, K. (2001) 'Accessing modern science: Policy and institutional options for agricultural biotechnology in developing countries', *World Development*, vol 30 (6), pp931–948
Herring, R. (in press) 'Stealth seeds: Biosafety, bioproperty, biopolitics', *Journal of Development Studies*, vol 43 no 1
James, C. (2005) *Global Status of Commercialized Transgenic Crops: 2005*, International Service for the Acquisition of Agri-Biotech Applications (ISAAA), Manila
Louwaars, N. P., Tripp, R., Eaton, D., Henson-Apollonio, V., Hu, R., Mendoza, M., Muhhuku, F., Pal, S. and Wekundah, J. (2005) 'Impacts of strengthened intellectual property rights regimes on the plant breeding industry in

developing countries: A synthesis of five case studies', report commissioned by the World Bank, Waningen UR, Centre for Genetic Resources, Waningen, The Netherlands

Morse, S., Bennet, R., and Ismael, Y. (2005) 'Comparing the performance of official and unofficial genetically modified cotton in India', *AgBioForum*, vol 8 (1), pp1–6

Pray, C., Ramaswami, B., Huamg, J., Hu, R., Bengali, P. and Zhang, H. (2006) 'Costs and performance of biosafety regulations in India and China', *International Journal of Technology and Globalisation*, vol 2 (1/2), pp137–157

Raney, T. (2006) 'Economic impact of transgenic crops in developing countries', in Chua, H, H. and Tingey, S. V. (eds) *Current Opinion in Biotechnology, Themed Issue on Plant biotechnology*, www.sciencedirect.com

Scoones, I. (2006) *Science, Agriculture and the Politics of Policy: The Case of Biotechnology in India*, Orient Longman, Hyderabad

United Nations Development Programme (UNDP) (2001) *Human Development Report 2001: Making New Technologies Work for Human Development*, Oxford University Press, New York

The Role of Government Policy: For Growth, Sustainability and Equity

Sakiko Fukuda-Parr

Chapter 11 compared the evolution of the technology and institutions. This chapter focuses on the role of government policy in this evolution. Here I return to the issues that were raised in the introductory chapter to draw lessons for making use of biotechnology for development – for growth, equity and sustainability.

Government policy: Objectives

In Chapter 1, I argued that while much of the international debate about biotechnology in developing countries focuses on whether GM crops would help reduce poverty and hunger, this is not the only national priority for development; there are five categories of policy objectives to which GM crops could contribute:

- to accelerate growth of GDP and exports;
- to participate in the cutting edge of global science and technology and remain internationally competitive in agricultural markets;
- to increase farm incomes and well being and hence reduce poverty;
- to reduce hunger and improve food security;
- to decrease chemical use to reduce environmental and health hazards.

What emerges from the five case studies is the importance of the first two objectives, in the overall context of a national vision of building internationally competitive agriculture as a driving force behind government objectives. Alongside other competing driving forces are the opposition movements that have championed the other three objectives, with a strategy to arrest or slow down the adoption of GM crops. Government strategies for agricultural biotechnology are well articulated in some countries (China, South Africa, Brazil) and ambiguous in others (India), but reflected in budgetary allocations to biotechnology R&D and in biosafety legislation.

Global integration agenda: Technology and markets

China, Brazil, India and South Africa all have strong commitment to developing national science and technology capacity overall. Biotechnology is a major advance that can hardly be ignored for countries that aim to remain competitive in global markets in an age when technology is an important factor in productivity. Moreover, soy, maize and cotton are all major traded commodities with large export markets. Policy makers in exporting countries would naturally consider access to the latest global technologies to be an important national priority for maintaining or increasing their shares in global markets. For each of the other countries, the GM crop that diffused is in each case the crop that is the leading agricultural export commodity: cotton in China and India, maize in South Africa.

   Table 12.1 shows the exports of soy, cotton and maize from the five countries. Soy in Argentina and Brazil and cotton in India and China are important export products, representing over 22 per cent of all agricultural export earnings. The case is particularly obvious for soy in Argentina and Brazil. These

Table 12.1 *Importance of global maize, cotton and soy markets for Argentina, Brazil, China and India (exports of crop as % of agricultural exports and total exports, average 2002–2004)*

	Maize		Cotton		Soy	
	% ag exports	% total exports	% ag exports	% total exports	% ag exports	% total exports
Argentina	8.5	3.8	0.2	0.1	62.7	28.5
Brazil	2.0	0.6	1.8	0.5	42.4	12.0
China	7.1	0.3	40.0	1.5	2.3	0.1
India	1.8	0.2	22.6	2.2	9.5	0.9
South Africa*	5.7	0.5	1.0	0.1	0.2	<0.1
US	9.8	0.8	3.5	0.3	15.4	1.26

Note: *South Africa average 2003–2004.

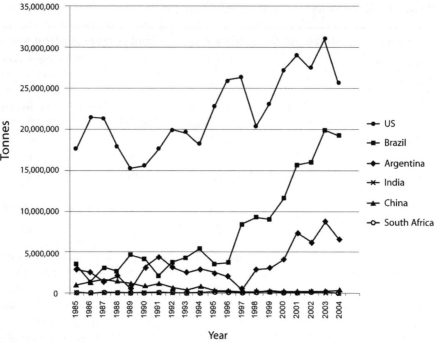

Figure 12.1 *Soy exports*
Source: Comtrade

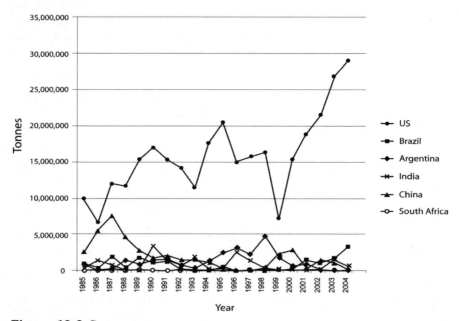

Figure 12.2 *Cotton exports*
Source: Comtrade

Table 12.2 *Comparing crop yields, 2004 (tonnes/ha)*

Country	Maize yield	Soy yield	Canola yield	Cotton yield
Argentina	6.4	2.2	1.3	0.4
Benin	1.1	0.5	NA	0.4
Brazil	3.4	2.3	1.7	1.0
China	5.1	1.8	1.8	1.0
France	9.0	2.5	3.5	NA
Germany	9.1	1.0	4.1	NA
Greece	10.1	2.1	NA	1.0
India	2.0	1.1	1.2	0.3
Mali	1.1	NA	NA	0.4
South Africa	3.1	1.6	NA	0.4
US	10.1	2.8	1.8	0.7

Source: FAOSTAT (http://faostat.fao.org) and Bulletin of the International Cotton Advisory Committee

countries rank second and third as exporters and soy products are important sources of foreign exchange for them; in 2004 soy accounted for about a third of all export earnings in Argentina and about 12 per cent in Brazil. In both countries, historical export levels of soy were stable up to 1995, fluctuating between 2 million tonnes and 4.5 million tonnes. Starting in 1997, after the introduction of GM soy varieties, expansion of production led to spectacular increases in exports – by 2004 to over 19 million tonnes in Brazil and 6.5 million tonnes in Argentina. (see Figure 12.1) In the case of cotton in China and India, exports have been falling. Though the crop is important for the domestic market and for rural incomes, global markets put pressure on policy makers to address the productivity of cotton production, particularly when yields per hectare are low compared with those of other major exporters, as is the case in India (see Figure 12.2).

Social equity agendas

The case studies refer to government policy commitment to social equity objectives. Brazil is concerned with the impact of GM crops on equity; South Africa made efforts to extend GM crops to the subsistence sector; China and India are focusing on major cash and staple crops grown by the family-farm sector; Argentina is using government revenues generated by soy exports to finance poverty programmes. However, reducing poverty and hunger are not the primary objectives, in contrast to the strategy defined in Central and West Africa. This does not necessarily mean that poverty reduction and social equity considerations do not receive attention in the five countries; rather, agricultural biotechnology is not considered to be a primary instrument for this purpose.

This is not surprising because the range of products already developed – the first generation of GM crops – are neither the crops nor the traits that are most important for poverty reduction and food security. Herbicide tolerance and insect resistant traits reduce costs and increase returns to labour rather than land and are thus suited for labour scarce environments. This is particularly the case for RR soy, which does not have significant yield benefits in most situations. *Bt* cotton, by contrast, does have significant yield benefits. The consensus view among economists and GM critics alike is that a pro-poor research agenda would target staple food crops, particularly those of the poorest people (cassava, millets, rice, wheat and white maize) and traits that increase production (yield potential, yield stability and environmental constraints such as salinity and drought) (Nuffield Council, 1999, 2004; Lipton, 2001; Naylor et al, 2004; FAO, 2004).

This pro-poor research agenda may evolve in the second generation of GM crop development. Brazil and China focus on priority food staples as other developing countries are doing in their public research[1] (Spielman et al, 2006). Brazil's policy is to leave the major 'platform crops' to the multinationals and to focus public sector efforts on locally important tropical crops, including cassava and beans. China's priority is rice, along with cotton; cotton is a major crop for small-scale family farmers. But yield increase does not appear as a major research priority of their biotechnology programmes in these areas (see Chapters 6–10).

India and South Africa are countries noted for their strong commitment to poverty reduction, and it is surprising that their governments have not articulated a pro-poor GM crop development policy. In India, political controversies over the technology have led to an ambiguous government policy. In South Africa, a series of government policy documents have explicitly identified science and technology, and biotechnology in particular, as a major tool of poverty reduction. Yet this has produced neither concrete products nor a concrete programme, probably as a result of many gaps in government policy. These gaps include credit and support services that led to the collapse of the *Bt* cotton experience in the Makhatini Flats. There also appears to be a lack of supply of well adapted varieties from R&D, as the *Bt* maize experience showed (see Chapter 10). That, in turn, is a reflection of weak public investment in development of commercially viable GM crops and a reliance on the private sector for plant breeding R&D, as discussed later in this chapter. Finally, because the majority of the poor in South Africa are in urban rather than rural areas, there may well be less public policy pressure to address productivity constraints in traditional agriculture as a priority in poverty reduction strategies.

What has been the distributional impact of GM crops so far? The extensive analysis of the Green Revolution points to two factors that are key in determining the distributional impact of a new agricultural technology: that the technology itself be scale neutral, and that the overall institutional context within which the technology is introduced not be biased against the poorer farmer, such as in access to credit and inputs (Hazell and Ramasamy, 1991). The economic analysis of on-farm benefits of GM crops carried out so far consistently shows that the technology itself is scale neutral, in many places performing better for small-scale farms than large-scale farms (FAO, 2004).[2] As for the context into which the technology is introduced, South Africa and Brazil are two countries with markedly dualistic agrarian structures in which huge commercial farms of thousands of hectares exist along with small-scale subsistence farms. In both countries, the new technology has done well in smaller scale farms, but in commercial rather than in subsistence contexts.

Another lesson of the Green Revolution is that the distributional consequences are felt beyond the farm level, through socio-economic changes such as demand for employment of landless labourers and ecosystem changes that take longer to be felt. Hypothetically, herbicide tolerant traits (RR soy, RR maize) can be distributionally biased against landless labourers as they reduce employment demands. But to date little empirical analysis has been done on these broader and longer term concerns. In Argentina, concerns have begun to be raised about the consequences of GM technology, and of soy cultivation in general, on concentration of land and on employment patterns (see Chapter 6). Much more research needs to be done on these issues in all countries.

The case studies also raise an important issue of impact on seed markets, where the larger joint ventures with multinationals have acquired large portions of the seed market, as in Brazil and India, squeezing out smaller local companies. Will this technology bring what the critics fear – a monopolistic control of the seed supply in the hands of multinationals and their affiliates?

Local contexts: National policy, stakeholders

GM critics often attribute the spread of GM crops to marketing and lobbying by powerful seed companies. For example, the anti-GM campaign, GMWatch, states prominently on its website, 'Waning confidence in GM in Europe, Australasia, Japan, and even parts of North America, has led the biotech industry to target developing Asian countries, where public understanding of the issue is still low. Massive hype and US trade pressure are being used to force in GM crops; … African countries are being targeted by the biotech industry and its lobbyists with unprecedented backing from the US government. Even food aid has been used to push GM into Africa' (GMWatch, 2006).

But what are the interests of the stakeholders in the developing countries, of local seed companies, research institutions, government officials, politicians, and last but not least, of the farmers themselves? Interests align for a pro-global integration agenda on the part of the local seed companies that see market incentives; research institutions that want to pursue the new science; and government officials in charge of technology, trade, agriculture, economy and finance. These stakeholders would see non-adoption as a risk of becoming marginalized from the global economy. The more important the crop is for foreign exchange, the stronger the pressures would be on finance, economy and agriculture officials. Non-adopting countries would compete with the US, the major exporter of these crops, and could lose market share and/or suffer from downward pressures on world prices.[3] These pro-global integration policy makers would ally with the country's scientists to promote the new technology. Finally, it is farmer demand that has been strongly felt; cotton farmers in India and soy farmers in Brazil have been the most prominent stakeholders.

Alternatively, where are the interest groups for a pro-poor agenda? There are farmers who would be most vulnerable to the ecological and socio-economic risks that have been raised. In both India and Brazil, indigenous people and others practising traditional agriculture are large constituencies; biodiversity is important to their livelihoods (Kageyama, 2005). Environmental groups in government, politics and civil society have thus pushed anti-GM strategies to ban or slow the spread of the technology altogether.

Government ministries in charge of poverty reduction, politicians and civil society advocates for development and social justice might be expected to see GM crops as an element of a poverty reduction strategy. But civil society groups dominate this debate and have in many instances allied with environment ministries to advocate banning the technology altogether rather than to develop national programmes with a pro-poor agenda. There has not been much momentum behind developing the third option of a new pro-poor GM strategy.

And what drives the anti-GM movements? They are led by anti-globalization and environmental groups such as Genewatch in India, or the Landless Movement in Brazil, which are linked with international networks and campaigns led by environmental and anti-globalization groups (Osgood, 2001). Some argue that economic interests are behind policy positions, at least in Europe; that economic interests of industry and agriculture that have 'lost out' to the US in biotechnology are mounting a protectionist move (Graff et al, 2004). Others emphasize the key role of public opinion behind European policies and socially embedded values about food and science (see Chapter 4; Osgood, 2001; Scoones, 2006) as well as a strategic concern with the growing power of corporations. Paarlberg (2001) points out that international NGOs and donor governments have been key influences from which anti-GM movements have drawn their support.

Government policy: Tools

Supportive government policy was instrumental in the development of GM technology in the private sector (see Chapter 3). This included:

- public financing and support to biotechnology in research institutions and universities, upstream of industry;
- a permissive biosafety process that is also relatively predictable for seed developers;

Table 12.3 *Comparing policy approaches*

	US	Argentina	Brazil	China	India	South Africa
Government support to biotechnology R&D (budget allocations)	Strong (increase)	Weak (decline)	Medium (stable)	Strong (increase)	Medium (increase)	Medium (stable)
Locus of R&D: i) upstream biotechnology;	i) Public and private (universities, research institutes, multi-national, biotech start ups)	i) Public national	i) Public national	i) Public	i) Public	i) Public
ii)biotech step of commercial product devt	ii)Multi-nationals (with some exceptions)	ii) Private multi-national	ii) Public, private, partnerships	ii) Public	ii) Public	ii) Public, private
iii)plant breeding step of commercial product devt	iii)Multi-nationals	iii) Private multi-national/local seed companies	iii) Private, public, partnerships	iii) Public, private multi-national and local	iii) Private multi-national and local seed companies	iii) Private
Biosafety legislation – precautionary or permissive[4]	Non-precautionary	Non-precautionary	Precautionary	Non-precautionary	Precautionary	Non-precautionary
Biosafety legislation – predictable implementation	Strong	Strong	Weak	Strong	Weak	Strong
Patents – UPOV signatory recognizing farmers' right to save seeds	No	Farmers' rights	Farmers' rights	Farmers' rights	Farmers' rights	Farmers' Rights
Patents – genes and plants	Yes	No	No	Yes	No	No
Patents – enforcement	Strong	Weak	Weak	Weak	Weak	Strong

- a strong IPR legislation that recognizes patents for genes and plants, and that is strictly enforced through the courts.

This policy environment facilitated US-based multinational corporations in developing varieties for US farmers and extending their seed markets internationally. How do the developing countries policies compare? Table 12.3 summarizes country approaches.

Support to R&D

As Chapter 11 documented, the strength of support to R&D, gauged by resource allocation, ranges from massive investment in China to strong commitment in Brazil, to an ambiguous position in India, to support in principle backed with meagre resources in South Africa, to almost no support in Argentina. These countries also differ markedly in the respective roles of the national public sector and private multinationals, particularly in the plant breeding stages. China, Brazil and India emphasize public roles through all the stages of R&D from upstream research to the plant breeding stages, while Argentina and South Africa leave the plant breeding to multinationals and seed companies.

These differences are not surprising in view of the broader national contexts of the five countries. In Argentina and South Africa, the overall domestic economic policy has emphasized liberalization and global integration. Introduction of GM varieties took place in the context of a major effort by the government to liberalize the economy and open up to foreign investment. In South Africa, agriculture had been highly protected for much of the 20th century, but began to be deregulated in the 1980s. The government's approach to promoting agricultural biotechnology appears to have been to set up enabling policies and centres, such as those promoting information for private investment, while government investment was focused on upstream biotechnology research and capacity building. Plant breeding was left to the private seed companies – as in the US. These incentives had little impact in provoking a response. Why? Perhaps there is inadequate market incentive in the system for private sector investment, or an overwhelming lack of capacity to use the technology in the private sector. Why did the entrepreneurial response that emerged in the Indian private seed sector not happen in South Africa?

These contrasting experiences show that in the absence of public investment, countries would be able to take advantage of the spillover benefits of GM global technology through either private–public partnerships or private joint ventures. But the private sector may well not see adequate market incentives to expanding the possibilities of using this technology further to tackle locally specific challenges. The response therefore would depend on the public sector. As the Argentina case study concludes, the dependence of Argentinian agriculture on R&D outside of the national public sector is a matter of concern. It leaves the country without a dynamic capacity to develop technology that responds to local needs.

Country studies in this volume, except the one for China, identify lack of financing and skilled resources as a major constraint to developing a more dynamic R&D system. But little is known about the real costs of developing GM crops. While often thought to be out of reach for most small developing countries, the West and Central Africa study proposes a project of US$19 million, well within line of major regional efforts.[5] There are also new ways for financing public R&D that could be explored in the context of private sector dominated activity. A portion of the technology fee, for example, could be used to finance a pro-poor research programme.

Another gap in the current policies of the five countries is collaboration outside of the public research institutions – with the private sector, and with public international and national institutions. Much of the literature on the shifting locus of R&D from the public to the private sectors emphasizes private–public partnership as the logical way forward for the development of agricultural biotechnology in developing countries. There are virtually no such experiences in the five countries, particularly in the biotechnology phase of commercial variety development. Brazil is the only case study that refers to this model. Otherwise, private–public partnership is limited to upstream biotechnology research as in South Africa. Private sector involvement is concentrated in private–private joint ventures with local seed companies. There is also, surprisingly, no mention of public–public collaboration either with international centres of the CGIAR system or among national research institutions of developing countries.

Biosafety legislation

Countries have been developing their biosafety legislation, revising laws as they gain experience. All countries have adopted legislation and application procedures that are more restrictive than those of the US, where transgenic crops are treated the same as conventionally bred crops. Whereas Brazil and India have adopted legislation based on precautionary principles (see Chapters 7 and 9; Paarlberg, 2001), China, South Africa and Argentina have not. Observers have noted that the implementation of the approval process is more predictable in China and South Africa than in India, where the process is mired in bureaucratic demands that can be costly in terms of both finance and time (Pray et al, 2006). Restrictive implementation makes the process unpredictable for the seed supplier attempting to obtain approval.

Have precautionary approaches and restrictive implementation led to higher standards of biosafety stewardship? Ironically, the impact may be perverse as it has merely led to the proliferation of the 'informal sector', which has brought supplies of unknown quality and been difficult to control. China recognized the obstacles to enforcement so its approach was to encourage these suppliers to go through the approval process.

By contrast, biosafety legislation has had major consequences for the supply of GM varieties, for the pace of their development and spread, and for the cost of compliance with the process, increasing R&D costs. In China, the cost of the biosafety process is a major issue for the small provincial research stations. Ironically, too, the more obstacle laden process in India has led to a less competitive market in GM seeds – leaving the Monsanto joint venture with a monopoly of the approved *Bt* cotton supply. As Scoones (2006) has written in his latest analysis of India:

> All want effective regulation, all want control in the public interest, licensing, policing and punishments that work. ... And if regulations become stricter, the hurdles for release become even more difficult to cross and the conditions of licensing become even more stringent (as opposition groups hoping for fewer GM releases argue), then it is likely that the only people who will be able to jump through these hoops will be the large corporations and their affiliates. The small players, perhaps those best able to produce crops adapted to local needs, will be squeezed out by the cost of regulatory compliance. The net losers will be the poorer farmers, the very people who the opposition groups argue are their constituency.

Patents

Intellectual property systems seek to balance the public priority for stimulating investment in innovation by giving a temporary monopoly to inventors, and giving to the public priority for diffusing new inventions for the benefit of users. Intellectual property protection for plant varieties has evolved over the last century. Patenting of genes and GM plant varieties is recent, introduced in the US in 1980 with a Supreme Court ruling (Louwaars et al, 2005; Chan, 2006). Only a few other countries recognize patenting of genes and GM plant varieties. The five countries in this study have been strengthening their IPR legislation to be TRIPS compliant, but most include a proviso, as permitted under the TRIPS agreement, to exclude genes and plants from patenting (Louwaars et al, 2005). All five countries give weaker protection to plant breeders than the US; most countries recognize the right of farmers to save seeds for their own use under UPOV. But the full implication of these patent regimes is not yet clear, since much will depend on how the patent offices and courts interpret the legislation. Just as patent disputes have become a significant issue in global markets, they may well emerge in developing countries as local research becomes more productive.

Has weak intellectual property protection been a constraint to innovation? Huang and Pray note that this is a major challenge that policy makers face in China. Developing country considerations for balancing incentives to breeders and incentives for diffusion are quite different from those in the US context. Multinationals based in the US and elsewhere who are investing large sums in biotechnology and plant breeding need large markets for their products and strong patent protection on genes, as well as on tools and varieties, to protect their R&D investments. Developing countries can license these technologies. Can revenues from patent licenses finance R&D for small markets of developing country research?

As we know from the pharmaceutical sector, strong patents can be an incentive to develop high price products for high income consumers, but can do little to encourage investment in high need products for low income consumers. In the pharmaceutical sector, this has led to large investments in diseases of the wealthy and neglect of diseases of the poor – or 'orphan diseases'. Naylor et al (2004) have argued that a similar process could be at work with private investments in agriculture. In this situation, developing country priorities for R&D will not likely be adequately financed from the private sector, even taking account of initiatives for public–private partnerships and philanthropy because of the limitations of financial incentives and other obstacles such as liability (Spielman and Von Grebmer, 2004; Chan, 2006; Osgood, 2006; Spielman et al, 2006). Will the public sector be the only source of financing – and in that case, what kind of patent protection would work best to stimulate research in public sector institutions for varieties with high social and low financial returns? Much more research is needed to explore the implications for developing countries. In this context, the practical difficulties of enforcing patents, especially for crops like soy, and the positive impacts of weak patents on seed prices, as well as on entrepreneurial response to develop adapted seed varieties for farmers, cannot be ignored.

GM crops for poverty reduction and food security: The path not yet taken?

What are the prospects for the 'gene revolution' forging a pro-poor path? In 1999, the seminal report on prospects for GM crops in developing countries by the Nuffield Council on Bioethics warned:

> As GM crop research is organized at present, the following worst-case scenario is all too likely; slow progress in those GM crops that enable poor countries to be self sufficient in food; advances directed at crop quality or management rather than drought tolerance or yield enhancement; emphasis on innovations that save labour costs (for example herbicide tolerance), rather than those that create employment; major yield-enhancing progress in developed countries to produce, or substitute for GM crops now imported in conventional (non-GM) form from poor countries.

The story of the first generation of GM crops closely resembles the worst-case scenario. This is clearly the product of the incentive structures of the corporate business model that supplies products that do not target and can bypass pro-poor priorities. But the emergence of the NARS-led model in China, as well as to a lesser extent in Brazil and India, that could compete with the corporate products, offers potential for a research agenda driven by national priorities rather than market incentives. The first generation products have been instrumental for developing countries in their global integration strategies. The second generation of GM crops, coming from China, Brazil and India, may well take a different path and help countries with domestic needs – and could also be targeted to benefit resource-poor farmers. It is also often thought that China, Brazil and India are exceptions so that the spread of GM technology will stop there. The experience of these countries, together with that of Argentina and South Africa, shows that technologically, developing country private and public sector entrepreneurship is able to respond rapidly to strong farmer demand for new seeds, taking advantage of the genetic work done elsewhere. But recent developments in China and India, with the approval of hundreds of events in China and the release of *Bt* cotton varieties using genes from India and China, also show the vibrant capacity of these NARS. Yet the diffusion of these crops to countries that lack the super NARS capacity in biotechnology and agricultural R&D needs more public policy facilitation to give incentives to developing country seed companies to market internationally.

Besides financing R&D, international public support is needed to develop new institutional approaches and policies. Farmers need a seed market that can deliver improved varieties that respond to their local needs, at low cost. By far the most complex obstacle has been to create a regulated market. But the regimen of strict patents, as in the US, or the stringent biosafety standards based on precautionary principles, as in Europe, have perverse effects, creating neither incentives for private investment nor higher standards of protecting the environment. This is the current model that has been assumed to be a prerequisite for developing country access to agricultural biotechnology, and that has driven the policy reforms that are commonly advocated as a means to harnessing this technology for poverty reduction (Graff et al, 2005). What should be the institutional path to creating such markets? Much more research is needed on many issues:

1   the longer term socio-economic structural consequences that this new technology can bring, to patterns of landownership and to employment;
2   the longer term socio-economic consequences of the structural shifts in seed markets, including the tendency for larger companies to wipe out smaller local companies;
3   the consequences of alternative models of R&D policy, the appropriate role of national public sector research and reliance on multinationals and seed companies, and the real potential for private–public partnerships;
4   the type of patent regimes and biosafety regimes needed to create a regulated but dynamic seed market capable of delivering better performing varieties to farmers at low prices, including alternative mechanisms for enforcing patents, such as at marketing points;

5   the levels and sources of financing required to support a pro-poor agenda. Little is known about the real costs of developing GM crops;
6   the real obstacles and opportunities for private–public partnerships;
7   the role of international investment in global technology in the pro-poor GM agenda. The current allocation of US$24 million to the CGIAR is dwarfed by comparing it to what the private sector and large NARS are investing.

But what is missing in the GM technology development story so far is proactive pro-poor support from the international public sector (World Bank, 2004),[6] global public support for a pro-poor R&D agenda, a political alliance of pro-poor civil society advocacy for mobilizing new technology for human development, and public financing for developing country access. Without such an alliance, a new R&D agenda focused not on export crops but on crops most important for poor farmers and poor consumers is not likely to emerge. But for that kind of alliance to emerge, a new social dialogue is needed that can break the gridlock driven by opposition that cannot be by-passed or ignored. The opposition is driven by a mistrust of science and fundamental questions about the nature of agriculture in society: what food? What farming? What society? These questions call for a new approach to democratic debates about biotechnology in the 21st century, and to forging new policies that would combat technological divides between developed and developing countries, between large resource-rich and small resource-poor developing countries, and between poor farmers and agribusiness within countries.

Notes

1   Spielman et al (2006) note that an analysis of all transformation events achieved by public institutions was done on the 15 crops considered critical to achieving food security and reducing poverty by the CGIAR.
2   FAO (2004) provides an excellent survey of findings to date.
3   See economic studies that estimate losses for non-adopters, for example Qaim and Traxler (2005) and FAO (2004).
4   See Paarlberg (2001) for a comprehensive review and typology of government policy approaches ranging from 'promotional' to 'permissive', to 'precautionary' to 'restrictive'.
5   Discussions of the country studies for this volume at the Bellagio conference in May 2005 referred to a total of US$10 million to set up a laboratory infrastructure.
6   The 2004 evaluation of the CGIAR system concludes that the system has not responded adequately to the biotechnology revolution in agricultural science. The report as a whole is critical of the weak focus of the system on the core mission of increasing agricultural productivity. See

www.worldbank.org/ieg/gppp/case_studies/agriculture_environment/
cgiar.html.

References

Chan, H. P. (2006) 'International patent behavior of nine major agricultural
biotechnology firms', *AgBioForum*, vol 9 (1), available at www.agbioforum.
org/v9n1/v9n1a07-chan.htm

FAO (2004) *State of Food and Agriculture 2003–04. Agricultural Biotechnology: Meeting
the Needs of the Poor?*, FAO, Rome

GMWatch (2006) 'Focus on Africa', www.gmwatch.org

Graff, G., Roland-Holst, D. and Zilberman, D. (2005) 'Biotechnology and poverty reduction in developing countries', *Research Paper No 2005/27*, United
Nations University, WIDER, Helsinki

Hazell, P. B. R. and Ramasamy, C. (eds) (1991) *The Green Revolution Reconsidered:
The Impact of High-yielding Rice Varieties in South India*, Johns Hopkins Press for
IFPRI, Baltimore, Maryland

Kageyama, P. (2005) 'Comments on the lecture of Jose Maria da Silveira, UNI-
CAP in the seminar: "Socio-economic issues of agricultural biotechnology
in developing countries", Bellagio, Italy, 30 May–4 June, 2005', Ministry of
Environment, Brazil

Lipton, M. (2001) 'Reviving global poverty reduction: What role for genetically
modified plants?', *Journal of International Development*, vol 13, pp823–846

Louwaars, N. P., Tripp, R., Eaton, D., Henson-Apollonio, V., Hu, R., Mendoza,
M., Muhhuku, F., Pal, S. and Wekundah, J. (2005) 'Impacts of strengthened
intellectual property rights regimes on the plant breeding industry in developing countries: A synthesis of five case studies', report commissioned by
the World Bank, Waningen UR, Centre for Genetic Resources, Waningen,
The Netherlands

Naylor, R. L., Falcon, W. P., Goodman, R. M., Jahn, M. M., Sengooba, T.,
Tefera, H. and Nelson, R. J. (2004) 'Biotechnology in the developing world:
A case for increased investments in orphan crops', *Food Policy*, vol 29, pp15–
44

Nuffield Council (1999) *Genetically Modified Crops: The Ethical and Social Issues*,
Nuffield Council on Bioethics, London

Nuffield Council (2004) *The Use of Genetically Modified Crops in Developing Countries: A Follow up Discussion Paper*, Nuffield Council on Bioethics, London

Osgood, D. (2001) 'Dig it up: Global civil society's response to the plant biotechnology agenda', in Anheir, H., Glasius, M. and Kaldor, M (eds) *Global
Civil Society, 2001*, Oxford University Press, Oxford, pp79–107

Osgood, D. (2006) 'Living the promise? The role of the private sector in enabling small-scale farmers to benefit from agro-biotech', *International Journal
of Technology and Globalisation*, vol 2 (1/2), pp30–45

Paarlberg, R. (2001) *Politics of Precaution*, Johns Hopkins University Press, Baltimore, MD

Pray, C., Ramaswami, B., Huang, J., Hu, R., Bengali, P. and Zhang, H. (2006) 'Costs and enforcement of biosafety regulations in India and China', *International Journal of Technology and Globalisation*, in press

Qaim, M. and Traxler, G. (2005) 'Roundup Ready soybeans in Argentina: Farm level and aggregate welfare effects', *Agricultural Economics*, vol 32, pp73–86

Scoones, I. (2006) *Science, Agriculture and the Politics of Policy: The Case of Biotechnology in India*, OUP, Delhi

Spielman, D. J. and Von Grebmer, K. (2004) 'Public-private partnerships in agricultural research: An analysis of challenges facing industry and the Consultative Group on International Agricultural Research', *EPTD Discussion Paper 113*, IFPRI, Washington DC

Spielman, D. J., Cohen, J. I. and Zambrano, P. (2006) 'Will agbiotech applications reach marginalized farmers? Evidence from developing countries', *AgBioForum*, vol 9 (1), available at www.agbioforum.org/v9n1/v9n103-spielman.htm

World Bank (2004) *The CGIAR at 31: A Meta-evaluation of the Consultative Group on International Agricultural Research*, Operations Evaluation Department, World Bank, Washington, DC

Index

Join our
online community
and help us save paper and postage!

www.earthscan.co.uk

By joining the Earthscan website, our readers can benefit from a range of exciting new services and exclusive offers. You can also receive e-alerts and e-newsletters packed with information about our new books, forthcoming events, special offers, invitations to book launches, discussion forums and membership news. Help us to reduce our environmental impact by joining the Earthscan online community!

How? – Become a member in seconds!

>> Simply visit **www.earthscan.co.uk** and add your name and email address to the sign-up box in the top left of the screen – You're now a member!

>> With your new member's page, you can subscribe to our monthly **e-newsletter** and/or choose **e-alerts** in your chosen subjects of interest – you control the amount of mail you receive and can unsubscribe yourself

Why? – Membership benefits

✔ Membership is free!

✔ 10% discount on all books online

✔ Receive invitations to high-profile book launch events at the BT Tower, London Review of Books Bookshop, the Africa Centre and other exciting venues

✔ Receive e-newsletters and e-alerts delivered directly to your inbox, keeping you informed but not costing the Earth – you can also forward to friends and colleagues

✔ Create your own discussion topics and get engaged in online debates taking place in our new online Forum

✔ Receive special offers on our books as well as on products and services from our partners such as _The Ecologist_, _The Civic Trust_ and more

✔ Academics – request inspection copies

✔ Journalists – subscribe to advance information e-alerts on upcoming titles and reply to receive a press copy upon publication – write to info@earthscan.co.uk for more information about this service

✔ Authors – keep up to date with the latest publications in your field

✔ NGOs – open an NGO Account with us and qualify for special discounts

Join now?
Join Earthscan now!
name
surname
email address

Earthscan Member
[Your name]

Click to Change

My profile
My forum
My bookmarks
All my pages

www.earthscan.co.uk

The Natural Advantage of Nations
Business Opportunities, Innovation and Governance in the 21st Century

Edited by Karlson 'Charlie' Hargroves and Michael Harrison Smith

Forewords by Amory B. Lovins, L Hunter Lovins, William McDonough, Michael Fairbanks and Alan AtKisson

'I am particularly pleased with the new book, The Natural Advantage of Nations, which will, in effect, follow on from Natural Capitalism, and bring in newer evidence from around the world'
AMORY B. LOVINS, Rocky Mountain Institute

'This is world-leading work, the team deserves the loudest acclamation possible'
BARRY GREAR AO, World Federation of Engineering Organisations

'A seminal book, a truly world changing book... As part of the process of pulling together the people whose ideas they wanted in the book, [the editors] have pulled together a whole movement'
L. HUNTER LOVINS, Natural Capitalism, Inc

This collection of inspiring work, based on solid academic and practical rigour, is an overview of the 21st century business case for sustainable development. It incorporates innovative technical, structural and social advances, and explores the role governance can play in both leading and underpinning business and communities in the shift towards a sustainable future.

The team from The Natural Edge Project have studied and incorporated key works from over 30 of the world's leaders in sustainability. The book is also supported by an extensive companion website. This work takes the lessons of competitive advantage theory and practice and combines them with the sustainability paradigm, in light of important developments in economics, innovation, business and governance over the last 30–50 years.

Far from being in conflict with economics and business practices, this book demonstrates how we can improve the well-being of society and the environment while driving innovation in an increasingly competitive world.

Hardback 1-84407-121-9 Published January 2005

HOW TO ORDER:

ONLINE www.earthscan.co.uk
CUSTOMER HOTLINE +44 (0)1256 302699
EMAIL book.orders@earthscan.co.uk